Globalization on the Line

Globalization on the Line

Culture, Capital, and Citizenship at U.S. Borders

Edited by

Claudia Sadowski-Smith

palgrave

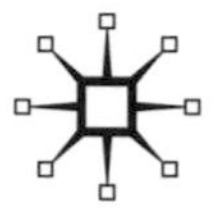

GLOBALIZATION ON THE LINE

First published 2002 by PALGRAVE™
175 Fifth Avenue, New York, N.Y.10010 and
Houndmills, Basingstoke, Hampshire RG21 6XS.
Companies and representatives throughout the world.

PALGRAVE is the new global publishing imprint of St. Martin's Press LLC Scholarly and Reference Division and Palgrave Publishers Ltd (formerly Macmillan Press Ltd).

ISBN 0–312–29482–4 hardback
ISBN 0–312–29483–2 paperback

Library of Congress Cataloging-in-Publication Data
Globalization on the line : culture, capital, and citizenship at U.S. borders / edited by
Claudia Sadowski-Smith
p. cm.
Includes bibliographical references and index.
ISBN 0–312–29482–4—ISBN 0–312–29483–2 (pbk.)
1. Globalization. 2. International economic relations.
3. Mexican-American Border Region. 4. United States—Foreign relations—1989- 5. United States—Cultural policy.
I. Sadowski-Smith, Claudia, 1968-

JZ1318.G679 2002
303.48'273071—dc21

2001056139

A catalogue record for this book is available from the British Library.

Design by Letra Libre, Inc.

First edition: June 2002
10 9 8 7 6 5 4 3 2 1

Printed in the United States of America.

To Bryan

Contents

Introduction

Border Studies, Diaspora, and Theories of Globalization[1]

Claudia Sadowski-Smith

Within the last few decades, U.S. borders have undergone tremendous change. Border regions have witnessed growing immigration and the relocation of industries under the North American Free Trade Agreement (NAFTA). NAFTA, one of the primary political instruments of globalization in the Americas, has, since its implementation in 1994, eradicated trade tariffs between Mexico, the United States, and Canada and thus rendered North American borders more porous to the free flow of goods and capital. But the agreement makes virtually no provisions for the free passage of people and has not prevented the further reinforcement of the U.S.-Mexico border. Here, since the 1990s, corrugated steel walls have begun to replace chain-link fences at the most popular crossing points. Giving primacy to unfettered movements of goods and investment capital, NAFTA has thus been working to create a common North American territory where goods and services can move more freely but where borders continue to intrude on the everyday lives of various groups of people.

It is within the context of this paradox that contributions to this collection examine a variety of cultural practices at U.S. borders where the effects of globalization have become especially visible. Bringing together artists, activists, and scholars from anthropology, Chicano studies, English, folklore, history, and political science, this volume locates the interdisciplinary cultural analysis of U.S. borders within debates about regional manifestations of globalization in the Americas, from which it has hitherto remained somewhat divorced. Such discussions are becoming especially important in the current context where plans for the extension of NAFTA into the whole Western Hemisphere are being finalized. The envisioned Free Trade Area of the Americas (FTAA) will constitute the world's largest supra-national free trade zone.

The collection sets itself apart from much of the academic scholarship and other influential discourses about U.S. borders that have tended to link change in the borderlands to contentious debates about the future of the U.S. American nation. Throughout the 1980s and 1990s, the U.S. media have repeatedly linked national security concerns to anxieties over cross-border violations by "illegal aliens" who supposedly undermine the "purity" of the U.S. nation. At the same time, academic work has developed an opposing view of the Southwestern border as a meeting place of diverse cultures and histories. This scholarship has emphasized cultural mixing and border fluidity as alternatives to state-sponsored forms of identity that have historically resulted in the denial of full citizenship rights to populations deemed undesirable and/or culturally and racially "other."[2] In their focus on immigration and cultural fusion, these sources of discursive influence, however, have less often considered other manifestations of change at the Southwestern border, such as the growing number of low-wage *maquiladora* (assembly) factories that produce consumer items for export.

At the beginning of the twenty-first century, issues of immigration across the U.S.-Mexico border no longer appear to be in the forefront of the national consciousness.[3] Despite some sporadic attention to economic and cultural fusion at the southern border during the past few years, the September 11, 2001 terrorist attacks on the World Trade Center and the Pentagon appear to have instead brought into sharper focus the physical border between the United States and Canada as a potential point of entry for terrorists. After a period of intense interest in the U.S.-Mexico borderlands and its hybrid border cultures in the 1980s and 1990s, cultural scholarship has recently also turned away from this location and moved *beyond* U.S. borders. Theories gleaned from the U.S.-Mexico border have become displaced by more abstract inquiries into diasporic notions of cultural identity and citizenship. Inspired by Paul Gilroy's explicitly diasporic perspective on black populations across the globe, the study of U.S. Chicana/os residents of *la frontera* as descendents of Mexican border crossers has expanded into analyses of transnational formations of U.S. Latina/os-'Latin' Americans.[4] Thus, in the early 1990s, Mexican scholar María Socorro Tabuenca Córdoba was still able to critique the largely figurative usage of the border as a synonym for Chicana/o identities as creating "a multicultural space in the United States [which] . . . erases geographical boundaries" (154). By the turn of the twenty-first century, however, the U.S.-Mexico frontier has evolved into one of the most prominent sites for analyses of border transgressions that emphasize contemporary diasporic practices of hybrid place-making and non-absolutist citizenship.

Within cultural critique, notions of "borders" and "border crossings" have, in fact, become synonymous with diasporic formations as border scholarship has paved the way for diaspora studies. Work on what anthropologist Arjun Appadarai has called "diasporic public spheres"—conceptualized in terms such as the Black Atlantic, the "Trans-Pacific," and the "U.S.-Mexico transfrontera zone"—has focused on experiences of displacement and cultural hybridity that link U.S. racialized groups to their Third World countries or areas of origin.[5] Members of diaspora exhibit multiple loyalties, move between regions, and often become themselves conduits for the increased flow of money, goods, information, images, and ideas across national boundaries. Their literal or symbolic forms of transborder movement undermine state-based nationalist ideologies and oppressive nation-state structures by defying a central aspect of state power—to define, discipline, control, and regulate all kinds of populations, whether in movement or in residence. The existence of diasporas has thus been centrally linked to arguments about the ongoing weakening of national borders and the demise of the nation-state as an object of emotional investment and as a major entity for the organization of social life and human activity.

Such assumptions about diasporas are greatly indebted to theories that have recently come under attack for positing cultural hybridity as a major challenge to structures of colonialism, neocolonialism, and imperialism.[6] But, perhaps more importantly, claims identifying diasporas as expressions of the nation-state's decline have also been informed by questionable premises about a fundamental antagonism between transnational processes and the nation-state. *Globalization on the Line* moves beyond this view of globalization and nation-states as two separate and opposed domains of theorization and politics, which has been essential to the neo-liberal, predictive rhetoric about global developments. Contributors to the collection focus on hemispheric cultural expressions of global processes that are currently reconstituting the relationship between nation-states and private corporations at U.S. borders. Essays in this volume point to the increasing multiplicity of actors in U.S. border zones and explore a wide variety of cultural expressions of citizenship that have emerged in response to domination by the U.S. nation-state and as a reaction to changes effected by globalization. The collection thus develops a view of U.S. borderlands as sites where conflicts between oppressive structures of the nation-state and globalization, on the one hand, and emerging alternative notions of societal membership, on the other, are currently being re-articulated in a variety of oppositional forms and strategies that encompass politically constructed affiliations and cross-cultural alliances.

After all, even though processes of globalization may largely be driven by the collaborative efforts of corporations and nation-states, they have also promoted the emergence of new global actors and the organization and internationalization of previously silenced groups. In the 1980s, social movements, organized around environmentalism, human rights, indigenous rights, and feminist causes, have intensified their efforts to work around nation-states by forging transnational lines of financial, cultural, and political support. Often manifested in complex efforts to build networks of transnational support, these groups' concepts of globalism differ from and/or tend to be directly opposed to the dominant, neo-liberal processes and rhetorics of globalization.

The term "globalization" itself has risen to prominence in the early 1990s to denote the emergence of more global patterns of social organization in the late twentieth and early twenty-first centuries.[7] As geographer Edward Soja has observed, while many features of the present moment have precedents in the longer histories of colonialism, imperialism, and capitalism (especially in nineteenth century free trade capitalism that abated after the world wars), "their intensification, interrelatedness and increasing scope makes present expressions different from the past" (20).[8] Divergent evaluations of global processes have, however, resulted in a variety of competing definitions of globalization. Depending on one's theoretical perspective, "globalization" can denote either the emergence of a more global consciousness or a collection of structural changes in the political, economic, and cultural realms; it can be understood as entailing the promise of liberation from national forms or it can embody the advent of new forms of transnational domination by private capital. Scholarship on U.S. borders has generally relied on a view of globalization as a process that erodes national frontiers and increases the potential to free marginalized ethnic groups from oppression by state and national forms. Having been most clearly articulated in cultural studies and anthropology, this stress on border porosity as a harbinger of global change and the accompanying tendency to overlook or minimize simultaneously occurring processes of border rigidity also characterizes other border scholarship. Here it often takes the form of an emphasis on the cultural, economic, and social blending of communities across borders that sets border zones apart from other regions in either neighboring country.[9]

The dichotomy between globalization and the nation-state has also been central to neo-liberal discourses of globalization, which are designed to promote minimal state intervention into the operations of private corporations. Such neo-liberal premises have significantly shaped the workings of global institutions, such as the World Trade Organization (WTO), the International Monetary Fund (IMF), and the World Bank, as well as free trade agreements like NAFTA. While it has increased corporate profits, sown the seeds of a Mexican middle class, and transformed Mexico into the second largest U.S. trading partner, NAFTA has also created conditions under which a variety of long-term negative developments have become more pronounced. In the absence of enforceable labor, human rights, and environmental protections, NAFTA has contributed to a rise in environmental pollution, accelerated de-industrialization and job loss in the United States and Canada, and promoted a sharp drop in Canadian currency. The agreement has also set the context for the emergence of far inferior working conditions in the assembly factories of Mexican border towns, which, in turn, have evolved into transit points for a growing number of northbound immigrants.[10] Market-driven processes of globalization as inscribed in NAFTA have thus not simply worked around and weakened national borders, but they have also helped to strengthen structural inequalities between the three participating countries and among various segments of their populations.

Some recent work has called for the separation of academic theory on globalization from the trappings of prevailing neo-liberal, predictive frameworks.[11] As part of her extensive work on global processes, sociologist Saskia Sassen, for example, has moved beyond an understanding of globalization that merely emphasizes the surge in transborder flow and the growth in internationalized labor, information, finance, and production networks. She has instead stressed changes in the organization and increasing interdependency of nation-states and transnational phenomena. According to Sassen, globalization entails the creation of a worldwide infrastructure that enables private corporations and financial, cultural, consumer, and labor markets to operate internationally. Thus, it is not so much the growth in global interconnectedness, but rather the strategic, precise, and focused nature under which conditions of globality have emerged at certain, privileged locations that radically distinguishes the present context from earlier forms of empire (Sassen "Globalization").[12]

These global conditions have been created through the cooperation of a variety of actors, among them corporations and nation-states, which are themselves changing. Joining a trade bloc, in fact, constitutes one of the most powerful mechanisms involved in altering relations between state and market, the government and private sector. On the one hand, states now increasingly act like corporations by encouraging domestic and international policies of deregulation, privatization, economic restructuring, and structural adjustment. As members of free trade agreements and global care-taking institutions like the WTO, nation-states have also abdicated their sovereignty over certain kinds of cross-border movements as well as given up their proactive role in using social policy and other programs for immediate national developmental needs. On the other hand, the neo-liberal removal of barriers to free trade has also hastened the transformation of multinational businesses into transnational corporations, which have redistributed their sites of production across various nation-states.[13] Similar to international trade and financial institutions that increasingly operate beyond the control of nation-states, corporations have acquired some of the powers of nation-states to exert control within a particular country and to shape its transnational politics, often by having the nation-state act on their behalf in the global arena.[14]

These new relationships between states and global processes have become especially manifested in certain geographies, such as U.S. border zones and urban areas like New York, Los Angeles, and Chicago, which have emerged as new scales of importance beyond the nation-state. As David Harvey has described the process by which certain geographies become inhabited by global (economic) forces in *Spaces of Hope,* capital finds some terrains easier to occupy than others, and some social formations insert themselves more aggressively into forms of market exchange than others (33). The global order inhabits national space unevenly and only partially with many differences across the geographical spectrum of the nation. Representing a specific set of social, cultural, and political relations, in geographer Henry Lefebvre's definition, geographical sites constitute both "products to be used" and also "means of production" (83). To put it differently, specific localities in the United States and elsewhere do not simply represent passive receptacles for flow and movement, but they also embody certain social, political, economic, and geographic conditions and units of agency that enable, shape, and sustain global change. As Saskia Sassen has pointed out, while they may also be present in other sites, new networks of global organizations and new types of political actors become more easily legible in the contexts of specific spatialities where connections between the global and the national are continually forged (Sassen "Nation States").

While few U.S. localities have remained completely unaffected by globalization, border areas and cities have become some of its foremost staging grounds. Academic scholarship on so-called global cities has produced a remarkable set of spatially-grounded perspectives on the relationship between postmodern urban sites and processes of globalization. This work has perhaps been most succinctly summarized in Edward Soja's theories of the "postmetropolis."[15] Soja writes that the postmetropolis is characterized not only by the emergence of new urban forms and the greater diversity of urban residents, but also by the surge in social inequalities among these populations. Drawing on Mike Davis's work in *City of Quartz,* among others, Soja's analysis also advances explanatory models for the continuing coherence of cities, such as the re-consolidation of state power within a proliferating number of carceral institutions. As it has largely been modeled after Los Angeles—a city whose post-suburban sprawl now arguably extends into Mexico—work on global cities thus offers potential intersections with scholarship on U.S. borders in its focus on analyzing the multiple effects of global forces beyond the perhaps more obvious processes of cultural, political, and economic mixing.[16] In comparison to the relatively well-defined boundaries of U.S. cities, it is admittedly harder to establish where exactly border areas begin and end, and how far the impact of a "boundary line" reaches into the surrounding areas. In recognition of these methodological difficulties, essays in *Globalization on the Line* limit their analysis of border areas to the kinds of locations near actual borderlines that have been visibly affected by and shaped changes in the U.S. nation-state and in the nature of global processes.

Several contributions add the U.S.-Canada frontier to a hemispheric scholarly framework that was originally spawned by imagining new forms of citizenship within U.S.-Mexico transborder cultures.[17] In establishing this long overdue comparative perspective, contributors nevertheless realize that it is difficult to speak about any border area in a singular way. While the 2,000 miles of U.S.-Mexico border are extremely diverse culturally, geographically, and economically, the 5,000 miles of U.S.-Canada frontier constitute a perhaps even less continuous zone since the presence of the culturally distinct Quebec interrupts any uniform interaction that might be postulated. *La frontera* crosses over forty-nine rural and urban U.S. border counties in Texas, New Mexico, Arizona, and California and over thirty-six rural and urban Mexican *municipios* (municipalities) in Tamaulipas, Nuevo León, Coahuila, Chihuahua, Sonora, and Baja California (Martínez 41). The border is made up of a variety of geographies and marked by increasing interactions between several, very different twin border towns, such as San Diego, California-Tijuana, Baja California; El Paso, Texas-Ciudad Juárez, Chihuahua; and Nogales, Arizona-Nogales, Sonora. Twice the length of the U.S.-Mexico border, the Canadian frontier is more sparsely populated and contains fewer urban mixing zones so that cross-border integration does not predominantly take place through border cities (Konrad 6). Yet there are a few historically developed border regions, including New England, the Great Lakes, the Prairies, the Rocky Mountains, the Pacific Northwest, and the Yukon River as well as urban border areas at Detroit, Michigan-Windsor, Ontario; Buffalo/Niagara Falls, New York-Niagara Falls/Toronto, Ontario; and the so-called Cascadia region that spans the Pacific Coast from mid-British Columbia south through Oregon and centers on Seattle, Washington and Vancouver, British Columbia.[18]

By including scholarship on both border areas, this volume addresses not only the lack of comparative work on U.S. borders, but also the relative dearth of work on the northern border's relationship to processes of globalization.[19] This oversight has most likely been the result of the strikingly different view on U.S. borders both in popular lore and academic scholarship. To a country preoccupied with churning out discourses that stress the flow of undocumented people (and of drugs) across the southern U.S. border, the

U.S.-Canada border has been of less concern. While the U.S.-Mexico border has been seen as the major battleground of U.S. citizenship in crisis, the line dividing the United States and Canada has, until recently, been touted as what is often referred to as "the world's longest undefended border."[20] As such, it does not appear to be a barrier to U.S. Americans, but seems only important in the context of Canadian efforts to resist the intrusion of a stronger and economically more powerful culture and of their attempts to define their own sense of nationhood.[21] Moreover, in contrast to Mexico and U.S.-Mexico border cultures, Canada's majority population appears to be ethnically similar to the majority of U.S. society. Even the slowly emerging recognition of Canada as a diverse and multicultural society that very much resembles the contemporary United States seems to be poised to further underscore the impression of the fundamental likeness of both countries.

The U.S.-Canada border has, however, also been recognizably transfigured by free trade, growing immigration, and by other processes associated with globalization. Discourses surrounding the terrorist attacks of September 11, 2001 that identified the U.S.-Canada border as the new "pathway for terrorists" into the United States and as yet another threat to U.S. national security interests were able to draw on fears linking a porous U.S.-Canada border to increases in undocumented immigration. In this regard, the rhetoric surrounding the U.S. northern border has come to resemble narratives about undocumented immigration that have defined the U.S.-Mexico border throughout the past few decades. The two U.S. borders are, however, not only connected to each other in terms of immigration, but also in terms of free trade effects on border communities. So far, the benefits of open trade have not been equally shared among the three participating countries or among various segments of the three countries' populations. While the NAFTA agreement does not prohibit the United States from employing trade policy, tax concession, or state subsidies to restrict Canadian and Mexican access to the U.S. market, for example, Mexican and Canadian authorities cannot do the same without being penalized by the United States (Drache 38). NAFTA thus illustrates Noam Chomsky's assertion that nation-states as sites of political, economic, and cultural sovereignty have not all been to the same degree reconfigured by transnational phenomena. Whereas Mexico (and partly Canada) have opened their economies and lifted restrictions on foreign investment under pressure from NAFTA, the United States continues to undermine the kinds of neo-liberal ideologies to which it ostensibly subscribes by instituting various forms of market protectionism for its own economy, while relentlessly bombarding other nations, including its neighbors, with the doctrines of free trade.

The two weaker economies have reaped fewer benefits from NAFTA in terms of increased market share or job creation. NAFTA has foremost created dependent sites of production in Mexican border areas, which mirror existing U.S.-Canada formations that first emerged in the early twentieth century (Hanson 42). Free trade has thus predominantly intensified the development of *one* of Mexico's economies (located in border towns), while the country as a whole is still faced with chronic unemployment, with millions subsisting on part-time work or struggling in the "informal" economy, and with wages declining in absolute and relative terms. As an unequal partner in NAFTA, Canada has also had to make a variety of concessions. The sharp decline in Canadian currency since the 1990s and the speed with which this depreciation has occurred has meant a decline in living standards for Canadian consumers who are lagging behind U.S. Americans in terms of affluence. The currency imbalance has also sped up the concentration of production, employment, and capital in Canadian localities that are generally situated close to the U.S. border, with U.S. border towns suffering from the decline in cross-border retail spending and the loss of jobs. While many Canadian companies have followed the lead of their U.S. counterparts in shifting manufacturing to Mexico, some Canadian and U.S. companies have also moved their production facilities to Canadian border zones, where costs are lower because of the cheaper Canadian dollar (Williams C5). The free trade agreement has thus formalized Mexico's economic dependency on the United States and reinscribed Canada's status as a "middle power" with a degree of direct foreign (U.S.) ownership unparalleled anywhere else on the globe (Panitch 82). Far from erasing national borders, NAFTA has rigidified economic, social, and political boundaries between the three countries and their populations.

Nothing could perhaps better illustrate the current economic imbalance between the United States and Canada than the recent decision of Kraft General Foods to relocate its Niagara Falls Nabisco plant, where shredded wheat has been produced for the past one hundred years, across the border to Niagara Falls, Ontario (Michelmore A10). The plant's present location has achieved symbolic status in Michael Moore's third movie *Canadian Bacon* (1994), which opens with a depiction of a closed Niagara Falls defense industry plant. The film explicitly links the end of the Cold War and the arrival of NAFTA to make yet another critical statement about the ongoing de-industrialization of the United States and the movement of many of the jobs lost in this process to locations abroad. At the time of this writing, there even exist plans for the creation of so-called Export Processing Zones (EPZs) in Canadian border zones. Providing the same kinds of conditions for foreign investment that characterize sites of *maquiladora* production in Mexican border towns, these zones are designed to attract Southeast Asian and European manufacturing and processing companies by allowing them to manufacture, process, store, and export goods duty free (Murphy n. pag.).

This book is divided into three parts, each articulating a specific theoretical aspect of its effort to contextualize scholarship on the two U.S. borders within processes of globalization. In the first section, contributors examine cross-border shopping, transnational production, and border migration along the U.S. borders with Mexico and Canada in order to place academic theories of change at the U.S.-Mexico border in dialogue with discourses about globalization. The second group of essays critiques tenets of diaspora studies, a field deeply indebted to analyses of the U.S.-Mexico border, by examining diasporic communities at the U.S.-Canada frontier and media discourses about transnationality at the U.S.-Mexico border. Contributors to the final section explore alternative globalisms in the U.S. borderlands. These essays chronicle or themselves undertake efforts at transnational coalition building and at the formation of transnational ties. Focusing on NAFTA-displaced women workers at the Texas frontier, squatter communities in Tecate, Baja California, and an environmental group at the Arizona-Sonora border, these articles make diverse communities at U.S. borders visible to one another and to structures of power embodied in the nation-state and in various processes of globalization. Collectively, contributors to the collection thus ponder the possibilities of a new kind of politics in response to the increasing transnationalization of the U.S. nation-state as it manifests itself at its borders.

Border Theories and Global Capital

Essays in the first section examine the relationship between the study of U.S. borders and several other disciplines—such as American and Chicana/o studies, diaspora and critical theory—and place them in dialogue with global processes. The prevailing emphasis on de-nationalization within border and diaspora studies can, at least in part, be understood as a consequence of the shift from the classic Marxist emphasis on capitalist production toward an analysis of various forms of nation-state based institutional power and of processes of marketing and consumption.[22] This shift theoretically mirrors the movement of First World capitalism toward service, information, and finance industries and away from traditional commodities and mass production since the latter part of the twentieth century. The more recent focus on transnational cultures as a means of liberation from normative national myths and restrictive state mechanisms has similarly remained caught in some of the neo-liberal hyperboles. Their endorsement of globalism as a predictive framework of evolution is based on the view of globalization as a force that affords the release of capitalism from its historical national containers. It thus perpetuates the classic liberal myth that capitalism constitutes an activity of private entrepreneurs freed from the interference of national machineries.

But the neo-liberal myth of minimal state intervention downplays the degree to which capitalism has historically been dependent upon nation-states to create the institutional conditions necessary for the existence of markets and to ensure accumulation, while legitimizing that process to workers and consumers (Hendrickson 15). Similarly, the wish to overcome state and nationalist repression through cultural transnational forces of globalization leaves unexamined the potential of nation-states to shelter their populations from periodic downturns of global business cycles, to submit investment and foreign policy decisions to public scrutiny and state regulation, to help preserve the existence of a public sphere, and to maintain public services, such as health care, education, and transportation, among others. Moreover, the general emphasis on border fluidity and the demise of the nation-state has also moved to the background considerations of the ongoing globalization of what Fredric Jameson has called structural inequalities between various parts of the world—nations, regions, and groups—which continue to articulate themselves on the model of national, ethnic, or diasporic identities rather than in other terms, such as those of social class. Jameson has defined this process as the transnationalization of "patterns of negative and positive exchanges which resemble those of class relations and struggles within the nation-state, even though . . . they do not (yet) define themselves in that way and remain fixed and thematized at the level of the spatial and geopolitical" (Jameson "Preface" xii).

Contributors to this volume emphasize such inequalities and also rescue from critical dismissal some of the potential roles of the nation-state. By focusing on the U.S. land border with Canada, Bryce Traister's opening essay reconfigures the currently dominant hemispheric perspective in American Studies to also acknowledge unequal relations between the United States and its northern neighbor. As he examines the phenomenon of U.S.-Canada border shopping, Traister points to similarities between post-national American studies and the corporate capitalist desire for permeable borders. He argues that both approaches are centered on the notion of a universalist subject, whose "hybridity"—devoid of any national difference—is wholly in the service of global market expansionism. Traister elevates the U.S.-Canada border to a position of new centrality for definitions of a decidedly anti-American national identity, which he reconfigures as an alternative to the current desire for geographical expansion beyond U.S. borders. This desire is equally manifested in the promotion of global consumerism and in the generatiion of new transnational perspectives in American Studies.

In his contribution to this section, Manuel Luis Martinez critiques border criticism as a set of methodologies within Chicana/o studies for its valorization of anti-nationalism and mobility over Mexican immigrants' struggles for community formation and an engagement with U.S. civil society. Martinez argues that border studies' celebration of Mexican immigrants' anti-statist "routes" reflects rather than contests the exploitative logic of both capitalism and of U.S. political and juridical practices of exploitation. As an alternative, Martinez examines the possibility of a revival of *Americanismo*—a paradigm of thought created by the Mexican American generation of the 1930s and 1940s to denote struggles for a stable communal space and civic participation in a more inclusive U.S. nation-state. Throughout his essay, Martinez analyzes texts by Ernesto Galarza, an important proponent of *Americanismo,* to show the dangerous proximity of what he calls postnational/poststructural forms of borderlands criticism to the capitalist logic of expansion that underlies current processes of globalization.

My own essay critically surveys border studies' identification of migration and diaspora as expressions of globalization that promote border fluidity and the nation-state's decline. I argue that this work's emphasis on the inter-ethnic coherence of contemporary diasporic groups as alternative expressions to national and statist forms of oppression tends to minimize cross-ethnic linkages based on (im)migrants' political identities. Examining one particular form of cross-ethnic transnational flow—the undocumented immigration of Mexicans and Chinese—the essay compares the effects of nineteenth century restrictive U.S. immigration laws with the ways in which current processes of globalization similarly constitute undocumented immigrants from the two countries. In an effort to build theories of affiliation below and beyond the level of diasporic formations, the article acknowledges parallels between poor and undocumented migrants from China and Mexico who are forced to negotiate increasingly complex conditions of border-free economics and border controls at the land borders with Mexico and Canada.

In the last essay of this section, Ursula Biemann analyzes another instance of the ongoing globalization of structural inequalities. Biemann juxtaposes the currently fashionable emphasis on the mobile and accelerated lifestyles available to certain First World subjects with the production of highly gendered, sexualized, and nationalized female subjects in the U.S.-Mexico border zone. Insisting on the materiality of the U.S.-Mexico border region, she also argues against the widespread tendency within cultural studies to treat the border region as a metaphor. Biemann focuses on structural relations in the U.S.-Mexico border zone by examining hitherto largely underexplored connections between the technological processes underlying *maquiladora* work and the militarization of the border, the smuggling of undocumented women immigrants to the United States, and the recent serial killings of *maquiladora* workers in Cuidad Juárez.

Border Communities and Theories of Diaspora

While essays in the first section of this volume characterize globalization at U.S. border sites as a collection of processes that is underscored by the effects of capital and market expansion, contributions to the second segment focus on theories of diaspora that have largely been gained from the analysis of U.S.-Mexico border cultures in literary/cultural studies and in anthropology. Scholarship on the history and multiplicity of diaspora that critiques the term's almost exclusive identification with current processes of globalization has only recently emerged.[23] Contributors to this section build upon such work but retain the volume's focus on manifestations of diasporic relationships across U.S. borders. While the essays by Claire Fox and Arlene Dávila critique relatively uniform assumptions about agency embodied in border studies scholarship, Donald Grinde's concluding article corrects its overemphasis on de-nationalization by highlighting the importance of place and national sovereignty for yet another group of border residents—indigenous borderlanders.

While contributors to this section do not deny the potential for resistance, they complicate the notion that the diasporic and the transnational can always be equated with politically progressive agency. Prevailing theories of cultural agency have emphasized how certain actors contest the hegemony of nation-states through heterogeneous practices of transnational citizenship that are conducted primarily in the realm of culture. The emphasis on individual agency within cultural studies initially emerged in reaction to the long-time dominance of structuralist Marxist claims for the importance of institutional, material, and ideological determinants. Cultural studies has criticized such a reduction of people to the ideological discourses or material conditions that constitute them and has instead stressed people's status as self-aware and self-determining subjects. While this approach recognizes that formal structures of ideology forge relations between social subjects and their conditions, it also emphasizes that people constitute individual (and often oppositional) meanings in response to their lived conditions by rearranging elements of dominant culture and ideology. This approach has, however, tended to overstate its own claims for change. Often assuming that all culturally heterogeneous practices imply opposition, this work has not sufficiently investigated how expressions of diaspora, rather than undermining nationalist and state-based forms of oppression, have themselves also remained shaped by the logic of nationalism and capitalism.

Claire Fox's article opens this section by focusing on historical antecedents of present-day transnational relations between the United States and Mexico as they have been created by cultural institutions rather than by the metaphorical or actual movements of people across borders. In her contribution, she traces the lingering effects of an older, binational configuration of mass media in the U.S.-Mexico borderlands. Developing an inter-medial lens, Fox examines the relationship of literary production to the hierarchically organized binational film cultures in the 1950s and 1960s—a time when the Mexico City-based film industry was in decline, while Hollywood remained strong. Rather than examining questions of individual agency by gauging the extent of literary resistance to the increasingly unequal organization of binational media at the time, however, Fox is interested in exploring how these cultural productions illuminate local conditions of audience reception. Her essay thus argues for a type of border studies that recognizes how representations of difference are always implicated in the uneven conditions of cultural production and consumption at both sides of the border.

Like Fox, Arlene Dávila argues that not all forms of transnationality are able to challenge normative hegemonic forms. Even though hers is the only contribution to the collection that does not explicitly focus on U.S. border areas but rather on medial movements across the U.S.-Mexico border, it is essential to a more differentiated understanding of contemporary expressions of diaspora. Dávila examines ongoing attempts by Hispanic media and marketing industries in both the United States and Latin America at forging a transnational, diasporic culture of *Latinidad.* She shows that their efforts to integrate U.S. Latinas/os by drawing on existing definitions of Latin American or U.S. Latina/o identity tend to reiterate essentialized and stereotypical discourses about cultural difference that have been ingrained in U.S. culture. In their homogenization of various diasporic populations, these attempts eradicate both intra-cultural and class distinctions. The Spanish-language media's notion of *Latinidad,* Dávila argues, not only reinscribes the racial logic of liberal multiculturalism at the level of the transnational nation, but it also emphasizes some other ingrained tenets of U.S. nationalism, especially the ideal of assimilationism to the U.S. ideal of upward class mobility.

While Dávila critiques media notions of *Latinidad* for their lack of attention to internal differences, Donald Grinde complicates the prevailing identification of borderlands transnationality with physical dispersal from a homeland. After all, the widely accepted view of Chicana/os-Latina/os as a diasporic people that reside in the borderlands of two nation-states resembles the self-understanding of indigenous border tribes who inhabited the Americas long before the consolidation of Canada, the United States, and Mexico as independent and separate nation-states.[24] But the growing literature on displaced communities has largely neglected to examine indigenous peoples, particularly cases where tribal identity is either tied to land that straddles the border or rooted in a community's separation by national borders. For example, some of the largest Native American reservations (e.g., of the Tohono O'odham) bridge the Arizona-Sonora border. Additionally, some indigenous borderlands tribes—such as the Kickapoo and Yaqui—have retained a strong sense of transnational tribal identity across the U.S.-Mexico border (Martínez 45–6).[25]

In his essay, Grinde focuses on the Iroquois Confederacy as one such binational borderlands community, which is, however, located at the U.S.-Canada border. Iroquois' struggles for sovereign nation status differ significantly from practices of diasporic place-making in that the Iroquois seek to retain their ties to land on both sides of the U.S.-Canada border. Citing various historical treaties as their legal authority, members of the Iroquois Confederacy move freely back and forth across this frontier and reconfirm their independence from both the U.S. and Canada. Grinde discusses Iroquois' struggles for the maintenance of border-crossing rights as symbolic of their historical resistance to oppression by the Canadian and U.S. nation-states and of their opposition to current processes of globalization. He chronicles historical and contemporary attempts at abrogating Iroquois' border-crossing rights, framed as nation-building efforts by both the United States and Canada. These were directed at eradicating indigenous notions of an American continent undivided by national borders. Grinde ends his essay by showing how nationalist strategies are today reinforced by increasingly more global corporate practices. Studying Iroquois' struggles for the maintenance of border crossing rights in the global context, Grinde's work thus complicates current debates of transnational citizenship and also expands scholarly notions of indigenous peoples as situated solely in local terms.

Border Alliances and Alternative Globalisms

In diversifying our notions of who counts as a border resident, Grinde's essay provides an excellent transition to the third section of this volume. In it, contributors examine how recent contradictory developments in border areas have been differentially negotiated by a heterogeneous group of border actors. Essays by artists and scholars in this section explore the transnational implications of the work of various borderlands groups, such as an organization of NAFTA-displaced women workers in El Paso, Texas; a collective of artists in San Diego, California; and a group of environmentally aware farmers in Sierra Vista, Arizona. Contributors ponder these groups' potential or actual ties with *maquiladora* workers in Ciudad Juárez, Chihuahua; a squatter community's struggles for self-determination in Tecate, Baja California; and the ongoing fight of the Zapatistas—the military arm of organized indigenous peoples in Chiapas—for environmental justice and land rights.

Essays in this segment thus show that new identities and forms of transnational citizenship are not only formed through institutions or through actual or metaphorical travels characteristic of diaspora, but also through processes of social mobilization. Contributors stress the rise in various cross-border and sometimes cross-ethnic affiliations that are organized around struggles for feminist objectives, human rights, and environmental concerns. These border activisms are not only shaped by resistance to exclusivist forms of nationalism, but have also emerged in opposition to inequities deepened by forces of globalization. Essays articulate diverse notions of transnational citizenship that need to be made visible, not only in order to legitimize some of the borderlanders' concerns, but also to contribute to articulations of incipient horizontal alliances as potential new forms of agency within a globalized context. To become truly global, some of these social movements will require coalitions among very different kinds of people with disparate goals and perceptions of the issues at hand.

Sharon Navarro's essay complicates the relatively monolithic image of a Latina/o-'Latin' American diaspora by revealing new complexities in the relationship between Mexican Americans and Mexicans. As she examines differential attitudes toward NAFTA by Mexican women working in the United States and in Mexico, Navarro moves beyond scholarly debates that have emphasized either the common interests among all people of Mexican descent or the various conflicts of interest between established Mexican Americans and newcomer Mexican immigrants. Navarro presents a case study of La Mujer Obrera, an organization of Mexican American displaced women workers in El Paso, Texas. La Mujer Obrera, the Working Woman, organizes Mexican American workers who have lost their jobs after low-wage, labor-intensive production moved into *maquiladoras* across the border, and at times literally into the adjacent border city of Ciudad Juárez. Navarro discusses the organization's contradictory cross-border relationship to Mexico as a result of efforts to re-integrate its members into a changing U.S. economy. While the organization has chosen not to establish solidarity networks with Mexican *maquiladora* workers and to restrict its activities to the national level, it has nevertheless appropriated elements of Mexican culture to frame appeals for support to U.S. state and federal institutions within diasporic notions of Mexican American identity.

Whereas Navarro's article ponders the surge of new economic and political boundaries separating Mexican women who live and work on both sides of the U.S.-Mexico border, Manuel Mancillas describes ways to traverse what he calls such newly emerging "labyrinths of corridors of power established by globalization." He chronicles the development of an ongoing collaborative project between the Border Arts Workshop/Taller de Arte Fronterizo (BAW/TAF) in San Diego/Tijuana and residents of the Poblado

Maclovio Rojas Márquez, a community of "squatters" in Tecate, Baja California, near Tijuana. Exploring the role of the artist/intellectual for cross-border community-building, Mancillas examines the conditions under which the BAW's public art program changed from an originally short-term, largely artistic endeavor into a long-term collaborative commitment. This transformation resulted from the artists' involvement in the Poblado's political struggles, which include wresting away control over the community's own fate from the Mexican state and combating the encroachment of transnational companies onto the Poblado's territory.

Emphasizing environmental concerns, Joni Adamson's article concludes *Globalization on the Line* by exploring potential connections between the struggles of indigenous Mexicans in Chiapas and the border activism by a group of ranchers near Sierra Vista, Arizona. Adamson shows that indigenous communities and the Malpai Borderlands Group alike stress the significance of local knowledges in the face of global organizations like NAFTA. As a small group of ranchers, the Malpai have developed environmentally-aware ranching practices—restoring rangeland, revolutionizing grazing practices, and conserving plant and animal life—as grassroots responses to cross-border liberalization that threatens their survival by favoring large corporate ranching enterprises. Adamson argues that the struggles of the Zapatistas and the activities of the Malpai group both directly confront mainstream definitions of "nature" as they have been inscribed into NAFTA. Adamson also notes that, like the Zapatistas, the Malpai have found ways to draw global attention to their situation by reaching out to larger social and environmental movements.

In their entirety, essays in *Globalization on the Line* re-evaluate the relationship of U.S. borders to processes of globalization. Emphasizing relations of symbiosis rather than antagonism between nation-states and transnational phenomena in the United States, contributors provide insights into the complexity of global processes as they take shape in U.S. border areas rather than reaffirm tenets of a neo-liberal, predictive framework that has postulated a rather uniform kind of globalization. This volume thus demonstrates that a rapidly advancing globalism requires new concepts of "borders," not only as designations for a variety of emergent notions of diasporic citizenship, but also as ways to think differently about alternatives to currently dominant neo-liberal accounts of globalization.

Notes

1. I would like to thank Claire F. Fox, Ann Ardis, and Jyotsna Singh for their careful reading of earlier drafts of this introduction. Thanks also to James D. Lilley for his assistance in the early stages of this project.

2. This border studies perspective relies heavily on Chicana/o studies work. Having emerged in the aftermath of civil rights struggles, Chicana/o studies originally promoted ideologies of cultural nationalism (*Chicanismo*), which tried to reclaim the long history of Mexican and Mexican American racialization in the U.S. (South)West by appealing to the pre-Columbian concept of Aztlán (the place from which the Mesoamerican Aztecs migrated to today's Mexico). For examples of influential cultural studies scholarship on borders, see the work of José David Saldívar and Héctor Calderón, Emily Hicks, Ramon Saldívar, as well as Saldívar's *The Dialectics of Our America.* On the importance of scholarship about the U.S.-Mexico border for the rethinking of the field of American Studies, see Carolyn Porter and Priscilla Wald.
3. Throughout 2000 and 2001, the mass media have sporadically paid attention to the U.S.-Mexico border by moving beyond (and sometimes also neglecting) processes of undocumented immigration to emphasize increases in other kinds of cross-border flows. This tendency may be witnessed in the 2000 release of the immensely popular film *Traffic,* a movie about the cross-border trafficking in drugs, and in a June 11, 2001 special issue of *Time* (entitled "Welcome to Amexica"). The *Time* issue examines various issues affecting the U.S. Southwest, such as the rise in Mexican drug cartels, the growing economies of Mexican border cities (especially Laredo and Ciudad Juárez), the development of a Latina/o-Hispanic culture in the United States, the emergence of U.S. American retirement communities in Mexican border towns, and the "Americanization" of Mexican politics. All of these stories were presented within a general framework stressing border erosion; the special issue was thus aptly subtitled "The border is vanishing before our eyes, creating a new world for all of us."
4. This is Daniel Mato's term. For the most influential scholarship exemplifying the cultural studies perspective on border areas, see José David Saldívar's *Border Matters.*
5. I have borrowed these terms describing various forms of diaspora from the work of Paul Gilroy, Rachel C. Lee, and José David Saldívar. Lisa Lowe's *Immigrant Acts* also begins the work of theorizing some of the similarities between a Mexican and an Asian (American) diaspora.
6. See Robert Young's *Colonial Desire,* which reveals similarities between eighteenth- and nineteenth-century discourses about dangerous racial contaminations and contemporary cultural theory's designation of "hybridity" as a term for cultural creativity and contestation. See also *Debating Cultural Hybridity* by anthropologists Pnina Werbner and Tari Modood, which draws attention to the early twentieth-century use of *mestizaje* (the ideology that affirms racial mixing between European settlers and indigenous people) in Latin America as a hegemonic tool against indigenous people. Their emphasis reinforces Young's point about the proximity of recent anti-hegemonic notions of hybridity to less liberating ideas. In *Hybridity and Its Discontent,* Avtar Brah and Annie E. Coombes, writing from a British context, have articulated a slightly different criticism of the way in which "hy-

bridity" has acquired the status of a common-sense term within cultural studies scholarship. These critics foreground the fact that hybridity, in its tendency to function as a descriptor of contemporary cultural processes, has often been represented as autonomous from any political and social determinations. They also point out that hybridity has mostly been theorized as a relation between colonized/diasporic communities and aspects of Western culture in ways that overlook how the internal dynamics of hybridity are constituted across and within social, political, and cultural entities. See also the recent collection of essays, entitled *Unforeseeable Americas Questioning Hybridity in the Americas* by Rita De Grandis and Zila Bernd that assembles work by Latin Americanists on the concept of hybridity. Of all these works, only Young's book explicitly links the terms "Hybridity and Diaspora" (in his first chapter) without in detail theorizing their connection.

7. For some of the most influential work on the nexus of culture and globalization see, for example, Fredric Jameson's *Postmodernism, or the Cultural Logic of Late Capitalism,* Robert Robertson, Malcolm Waters, David Harvey's *The Conditions of Postmodernity,* and Mike Featherstone. See also the more recent collection of essays by Fredric Jameson/Masao Miyoshi and Lisa Lowe/David Lloyd as well as Michael Hardt/Antonio Negri's work.
8. Debates emphasizing either disjunctures or continuities between present-day globalization and past processes have, by now, produced a relatively large body of work. See for example, Enrique Dussel in *The Cultures of Globalization.* With regard to the effects of capital and finance mobility on sovereignty in nineteenth-century free trade capitalism and neo-liberal free trade, Bill Maurer has recently argued that the very comparison wrongly presupposes the convertibility of different forms of property as the cornerstone of capital mobility and that it does not recognize distinctions in the objects of the movement—finance capital—and in the nature of the movement itself.
9. See, for example, the work of historian Oscar Martínez and of geographer Lawrence Herzog, which describes the U.S.-Mexico border as a place where people, cultures, and architectures are becoming more integrated to form a new culture that differs from that of either neighboring country. Herzog is a principal proponent of the view that global transformations in recent years have led to the declining importance of the U.S.-Mexico border. He claims that "the obvious change has been the shift from boundaries that are heavily protected and militarized to those that are more porous, permitting cross-border social and economic interaction" ("Changing Boundaries" 5–6). Recent press coverage has similarly turned to describing and celebrating instances of cultural and economic exchange across the border as manifestations of border erosion. A June 2000 CNN/*Time* co-production, "The New Frontier/La Nueva Frontera," for example, painted a picture of a region that is growing more closely together in such matters as employment, education, water use, and economic growth. A June 11, 2001 more in-depth companion piece *Time* article, entitled "A Whole New World," employs the reality of constant cross-border movements of goods and people (both legally and

illegally) to describe the border as a "barbed-wire paradox, half pried open, half bolted closed," but predicts that "if presidents . . . solve the problems of the two countries that need each other but don't completely trust each other, the American Century could give way to the Century of the Americas, and the border might as well have disappeared altogether" (Gibbs 38). The work of Thomas M. Wilson and Hastings Donnan provides a useful overview of social science work on the U.S.-Mexico border. For examples of scholarship on U.S. borders that makes the recognition of re-bordering central to its argument, see the work of David Spener and Kathleen Staudt as well as Saskia Sassen's *Globalization and Its Discontents.*

10. Some economists and journalists have recently stressed the fact that Mexican border towns are also beginning to develop other types of industries besides those geared toward export-oriented assembly. A 1998 *Business Week* article points to the rise in the exports of more sophisticated products that are manufactured in Mexico, including autobrake systems and laptop computers. Tim Padgett and Cathy Booth Thomas may have had some of the same examples in mind when they write about a Delphi Automotive Systems' Technical Center in Ciudad Juárez, where teams of Mexican researchers develop steering-column prototypes for U.S. car and thus perform the kinds of R&D that has tended to remain in First World countries.
11. See especially the work by Anna Tsing, Aihwa Ong, and Bill Maurer. Tsing asks scholars to give up the ingrained tools and frameworks used to discuss globalization, such as the distinction between "global" forces and "local" places (since all cultural processes are both socially and culturally particular and at the same time also productive of widely spreading interactions), assumptions that all new developments are always also universal or global, and efforts to homogenize distinctive cultural communities. Tsing argues that these prevailing approaches to globalization may have become so seductive and easy to work with because they resound with common sense in the United States.
12. Michael Hardt and Antoni Negri have termed this new form of global sovereignty "Empire." They define the current global system as being composed of a series of national and transnational organisms that are united under a single logic of rule. This system neither simply continues earlier forms of European colonialism nor does it mean the worldwide expansion of capitalism into areas that had not previously been penetrated, such as the former Soviet Union and its so-called satellite states. Instead, "empire" names the increasing de-centering of global power, the scrambling of the old territorial divisions into First, Second, and Third Worlds, and the transformation of productive processes to foreground communicative, cooperative, and affective forms of labor.
13. Anthropologist Roger Rouse has provided a detailed account of the shift from multi- to transnational corporations. He shows that after World War II, corporations began to use peripheral regions of the world as markets of raw material for domestic manufacturing and thus also stimulated the de-

velopment of mass consumer markets in Western Europe and selected Third World countries. With the loss of the U.S. global superpower status, however, these arrangements became subject to increasing stress. Especially within the last two decades of the twentieth century, multinational enterprises went transnational. While multinational corporations integrate self-contained production and marketing facilities in a number of different national sites, transnational corporations redistribute production across sites in different nations. Transnational corporations also encourage flexible forms of labor by employing both migrant labor and workers in Third (and also Second) World countries. Corporate-driven global economic restructuring has thus meant national economic deregulation and privatization in the North and more thorough forms of structural adjustment in the South.

14. Examples of the increasing imbrication of U.S. national politics and free trade abound. They include the formalization of provisions into free trade agreements that provide for "investor-to-state" lawsuits, which allow corporations to sue governments for compensation if they feel that any state action, including the enforcement of public health and safety laws, cuts into their profits. See also Bryce Traister's essay in this volume, in which he examines the ways in which the U.S. government has acted on behalf of corporations to ban the Canadian government's efforts of denying distribution rights to U.S. magazines with insufficient amount of Canadian content.
15. For work on U.S. global cities, see, for example, Janet L. Abu-Lughod, Saskia Sassen's *The Global City*, and James Holston.
16. Although urban spaces are now becoming increasingly transformed into "megalopolises"—assemblies of cities—whose outlines are being blurred as a result of urban sprawl, the effect of U.S. border lines' political, economic, and cultural inscription into the surrounding landscape is still less easily discernible.
17. The relationship between Canada, Mexico, and the United States has hitherto only rarely been treated comparatively. Most analyses look either at relations between the United States and Mexico (or Latin America) or, alternatively, at the U.S.-Canada nexus. Some hemispheric cultural studies (especially Asian American studies) have recently begun to compare the United States and Canada in terms of their similar (yet also very different) forms of multiculturalism by organizing their inquiries largely along ethnic lines. These studies often focus on the surge of ethnically constituted diasporic spaces as a result of ongoing Third World migration into the United States and Canada and stress the continuity of U.S. ethnic literature with cultural productions of Canada's racialized "visible minorities." For a general comparative perspective on the three countries in literary studies, see Smorkaloff; for scholarship on NAFTA's effect on the three countries' economy, see Drache; and for work from geography about the two land borders, see Konrad.
18. For more information on U.S.-Canadian border cities, see Peter Karl Kresl.
19. I would have liked to include more essays on the impact of globalization on the U.S.-Canada border into the collection, but could not locate such work

or find scholars willing to contribute. For a sociological perspective on the U.S.-Canada border and the rise in cross-border crime since the passage of NAFTA, see Ruth Jamieson et al. For a critical perspective on the much-touted regionalization of trade and economic relations in U.S.-Canada border regions, especially in the Pacific Northwest, see Matthew Sparke.

20. This denotation for the U.S.-Canada border is apparently widespread and has also been employed in a variety of scholarly publications, such as the article by Thomas F. McIlwraith. See also, for example, the cover of a 1995 video entitled *Adventure at the U.S.-Canadian Border,* which calls the Canadian frontier "the longest, continuous, undefended border in the world." U.S. Americans have very different views about enforcement at both borders. As Nancy Gibbs in a *Time* article puts it (somewhat incorrectly), the border with Canada "is basically defended by a couple of fire trucks, and most Americans think that's about all they need. The southern border is half as long, has the equivalent of an army division patrolling it, and many Americans say it should be buttoned down tighter" (39–40). According to a June 11, 2001 "*Time*/CNN Poll," 53 percent of U.S. respondents thought that it should become harder for Mexicans to cross the border into the United States as compared to 21 percent who thought the same about potential Canadian border crossers. Whereas 58 percent of the respondents believed that border regulations were about right for Canadians, only 25 percent had the same opinion of Mexicans. However, the small percentage (15 percent) of people who opined that it should become generally easier for people to cross U.S. borders thought so with regard to both Canadians and Mexicans ("*Time*/CNN Poll" 46).
21. The two borders have also undergone very different histories of militarization. While the U.S.-Mexico border was created in 1848 and only began to be supervised by U.S. Customs at the turn of the twentieth century, the U.S.-Canada border was established theoretically in the Treaty of 1818 but the boundary itself was enforced much earlier (LaDow 7). A variety of forts and war vessels were employed to militarize the border during the early nineteenth century in the wake of the wars of 1775 - 6 and 1812 - 14. After the 1871 Treaty of Washington initiated an era of peace between the two countries, the border became demilitarized and, in 1920, the two countries finally agreed upon a firm boundary line to be supervised by custom and immigration officials (McIlwraith 54).
22. Cultural studies has, since its inception in the 1970s in Great Britain, been broadly understood as an oppositional political undertaking aimed at raising concerns about social equity. Trying to mobilize members of subordinated groups into political activism against various forms of hegemony, the field has focused on examining heterogeneous cultural practices inspired by class, gender, sexuality, race, and ethnic affiliation. Initially conceptualized as a revision of certain forms of Marxism, cultural studies (especially in the United States) has more recently become dominated by post-Marxist approaches that insist on the relative autonomy of cultural practices from capitalist

modes of production and that tend to examine domination and hegemony in terms of the nation-state apparatus. For examples of British cultural studies scholarship, see Graeme Turner's work and the section on "Cultural Studies in Britain" in Jessica Munns and Gita Rajan.

23. See, for example, the work of Khachig Tölölyan and of Kandice Chuh.
24. Some work in border studies has already taken steps toward recognizing the concerns of indigenous borderlanders. Daniel Cooper Alarcón, for example, has pointed out that while it has selectively appropriated indigenous culture and mythologies, *Chicanismo* has also largely ignored the competing nationalist claims of indigenous people to the U.S. Southwest (Cooper Alarcón 9). Recent theories of intersections between the anticolonial struggles of Native Americans and Chicana/os-Latina/os are exemplified in Cherríe Moraga's work. In her latest collection of essays, *The Last Generation,* she proposes to combine the Chicano attempt at recreating the mythic Aztlán on the territory that until 1848 belonged to Mexico and pan-tribal Native American demands for the recovery of their land, while also acknowledging early indigenous Mexican resistance to colonization.
25. Living along the Texas border, the Kickapoo maintain ties with their home village Nacimiento in Coahuila, Mexico. The Yaqui of southern Arizona sustain connections with their parent Yaqui nation in Sonora. The Tohono O'odham have lived for centuries on desert that overlaps the Arizona-Sonora border, with 24,000 American O'odham occupying a 4,800-square-mile-reservation and 4,000 Mexican O'odham scattered in farms and towns on the Mexican side. Like the Kickapoo who for years maintained a way station known as Kickapoo Village under the international bridge connecting Eagle Pass, Texas and Piedras Negras, Coahuila, the Tohono O'odham possess their own crossing points in remote, sparsely settled locations (Martínez 45–6). For work specifically about the Yaqui and their complex relationship to the U.S.-Mexico border, see David Shorter.

Works Cited

Abu-Lughod, Janet L. *New York, Chicago, Los Angeles; America's Global Cities.* Minneapolis, MN: University of Minnesota Press, 1999.

Adventure along the U.S.-Canada Border. John Holod Production. Distributed by American Home Treasures, 1998.

Anzaldúa, Gloria. *Borderlands/La Frontera: The New Mestiza.* San Francisco, CA: Aunt Lute Books, 1987.

Appadurai, Arjun. *Modernity at Large: Cultural Dimensions of Globalization.* Minneapolis, MN: University of Minnesota Press, 1996.

Brah, Avtar and Annie E. Coombes, eds. *Hybridity and its Discontents: Politics, Science, Culture.* London: Routledge, 2000.

Canadian Bacon. Dir. Michael Moore. Perf. John Candy, Dan Aykroyd, and Rhea Perlman, 1994.

Chomsky, Noam. "Free Trade and Free Market: Pretense and Practice," *The Cultures of Globalization.* Eds. Fredric Jameson and Masao Miyoshi. Durham, NC: Duke University Press, 1998: 356–370.

Chuh, Kandice. "Transnationalism and Its Pasts," *Public Culture* 9.1 (1996): 93–114.

Cooper Alarcón, Daniel. *The Aztec Palimpsest: Mexico in the Modern Imagination.* Tucson, AZ: University of Arizona, 1997.

Córdoba, María-Socorro Tabuenca. "Viewing the Border: Perspectives from 'the Open Wound'," *Discourse* 18.1–2 (Fall/Winter 1995–96): 146–168.

Davis, Mike. *City of Quartz: Excavating the Future in Los Angeles.* London: Verso, 1990.

De Grandis, Rita and Zilá Bernd, eds. *Unforeseeable Americas: Questioning Cultural Hybridity in the Americas.* Amsterdam: Rodopi, 2000.

Drache, Daniel. "The Future of NAFTA in the Post-National Era," *Review of Radical Political Economics* 25.4 (1993): 30–44.

Dussel, Enrique. "Beyond Eurocentrism: The World-System and the Limits of Modernity," *The Cultures of Globalization.* Eds. Fredric Jameson and Masao Miyoshi. Durham, NC. Duke University Press, 1998: 3–31.

Featherstone, Mike. *Undoing Culture: Globalization, Postmodernism, and Identity.* London: Sage, 1995.

Gibbs, Nancy. "A Whole New World," *Time.* Special Issue "Welcome to Amexico" (June 11, 2001): 38–45.

Gilroy, Paul. *The Black Atlantic: Modernity and Double Consciousness.* London: Verso, 1993.

Hanson, Gordon H. "North American Economic Integration and Industry Location," *Oxford Review of Economic Policy* 14.2 (Summer 1998): 30–44.

Hardt, Michael and Antonio Negri. *Empire.* Cambridge, MA: Harvard University Press, 2000.

Harvey, David. *Spaces of Hope.* Berkeley, CA: University of California Press, 2000.

———. *The Condition of Postmodernity.* Oxford: Blackwell, 1989.

Hendrickson, Mary. "Post-Fordism and the NAFTA Debate," *Critical Sociology* 21.2 (1995): 7–17.

Herzog, Lawrence A. *From Aztec to High Tech: Architecture and Landscape Across the Mexico-United States Border.* Baltimore, MD: Johns Hopkins University Press, 1999.

———. "Changing Boundaries in the Americas: An Overview," *Changing Boundaries in the Americas.* Ed. Lawrence A. Herzog. La Jolla, CA: University of California at San Diego, 1992: 5–6.

Hicks, Emily D. *Border Writing: The Multidimensional Text.* Minneapolis, MN: University of Minnesota Press, 1991.

Holston, James, ed. *Cities and Citizenship.* Durham, NC: Duke University Press, 1999.

Jameson, Fredric. "Preface," *The Cultures of Globalization.* Eds. Fredric Jameson and Masao Miyoshi. Durham, NC: Duke University Press, 1998.

———. *Postmodernism, or The Cultural Logic of Late Capitalism.* Durham, NC: Duke University Press, 1991.

Jamieson, Ruth, Nigel South, and Ian Taylor. "Economic Liberalization and Cross-Border Crime: The North American Free Trade Area and Canada's Border with the U.S.A.," *International Journal of the Sociology of Law* 26.2 (1998): 245–272 and 26.3 (1998): 285–319.

Konrad, Victor. "The Borderlands of the United States and Canada in the Context of North American Development International," *Journal of Canadian Studies* (Fall 1991): 77–95.

Kresl, Peter Karl. *The Impact of Trade on Canadian-American Border Cities.* Orono, ME: Canadian-American Center at the University of Maine, 1993.

LaDow, Beth. *The Medicine Line: Life and Death on a North American Borderland.* New York: Routledge, 2001.

Lee, Rachel C. *The Americas of Asian American Literature.* Princeton, NJ: Princeton University Press, 1999.

Lefebvre, Henri. *The Production of Space.* Trans. Donald Nicholson-Smith. Oxford: Blackwell, 1991.

Lowe, Lisa and David Lloyd, eds. *The Politics of Culture in the Shadow of Capital.* Durham, NC: Duke University Press, 1997.

Lowe, Lisa. *Immigrant Acts: Asian American Cultural Politics.* Durham, NC: Duke University Press, 1996.

Martínez, Oscar. *Border People: Life and Society in the U.S.-Mexico Borderlands.* Tucson, AZ: University of Arizona Press, 1994.

Mato, Daniel. "On the Making of Transnational Identities in the Age of Globalization: The US Latina/o-'Latin' American Case," *Cultural Studies* 12.4 (1998): 598–621.

Maurer, Bill. "A Fish Story: Rethinking Globalization on Virgin Gorda, British Virgin Islands,"*American Ethnologist* 27.3 (2000): 670–701.

McIlwraith, Thomas F. "Transport in the Borderlands, 1763–1920," *Borderlands: Essays in Canadian-American Relations.* Ed. Robert Lecker. Toronto: ECW Press, 1991: 54–89.

Michelmore, Bill. "Nabisco's Exit to Be Felt on the Farm," *Buffalo News* (May 20, 2001): A-1 and A-10.

Moraga, Cherríe. *The Last Generation: Prose and Poetry.* Boston, MA: South End Press, 1993.

Munns, Jessica and Gita Rajan, eds. *A Cultural Studies Reader: History, Theory, Practice.* London: Longman, 1995.

Murphy, John. "Free Trade Zone in Saint John Bonanza at Whose Expense?" September 7, 1998. Available at http://www.afl.org/LabourNews/sept98–7.html Accessed January 14, 2001.

Ong, Aihwa. *Flexible Citizenship: The Cultural Logics of Transnationality.* Durham, NC: Duke University Press, 1999.

Padgett, Tim and Cathy Booth James. "Two Countries, One City," *Time* (June 11, 2001): 64–66.

Panitch, Leo. "Globalization, States, and Left Strategies," *Social Justice* 23.1–2 (Spring-Summer 1996): 79–90.

Porter, Carolyn. "What We Know that We Don't Know: Remapping American Literary Studies," *ALH* 6.3 (Fall 1994): 467–526.

Robertson, Roland. *Globalization.* London: Sage, 1992.
Rouse, Roger. "Thinking Through Transnationalism: Notes on the Cultural Politics of Class Relations in the Contemporary United States," *Public Culture* 7 (1995): 353–402.
Saldívar, José David and Héctor Calderón, eds. *Criticism in the Borderlands: Studies in Chicano Literature, Culture, and Ideology.* Durham, NC: Duke University Press, 1991.
Saldívar, José David. *Border Matters: Remapping American Cultural Studies.* Berkeley, CA: University of California Press, 1997.
———. *The Dialectics of Our America: Genealogy, Cultural Critique, and Literary History.* Durham, NC: Duke University Press, 1991.
Saldívar, Ramon. *Chicano Narrative: The Dialectics of Difference.* Madison, WI: University of Wisconsin Press, 1990.
Sassen, Saskia. "Nation States and Global Cities," Keynote Presentation at Michigan State University as part of the conference "Globalicities," October 19, 2001.
———. "Globalization: Developing a Field for Research and Teaching," Presentation at Dartmouth College as part of the conference "Globalization of the Academy," November 15, 2000.
———. *Globalization and Its Discontents.* New York: New Press, 1998.
———. *The Global City: New York, London, Tokyo.* Princeton, NJ: Princeton University Press, 1991.
Shorter, David. "Crossing National Borders: Indigeneity and Religion in the U.S.-Mexico Borderlands," Presentation at the annual meeting of the American Studies Association, Washington, D.C., November 8–11, 2001.
Smorkaloff, Pamela Maria. "Shifting Borders, Free Trade, and Frontier Narratives: U.S., Canada, and Mexico," *American Literary History* 6.1 (Spring 1994): 88–102.
Soja, Edward W. "Six Discourses on the Postmetropolis," *Imagining Cities: Scripts, Signs, Memory.* Eds. Sallie Westwood and John Williams. London: Routledge, 1997.
Sparke, Matthew. "Excavating the Future in Cascadia: Geoeconomics and the Imagined Geographies of a Cross-Border Region," *BC Studies* 127 (Autumn 2000): 5–44.
Spener, David and Kathleen Staudt, eds. *The U.S.-Mexico Border: Transcending Divisions, Contesting Identities.* Boulder, CO: Lynne Riemer Publications, 1988.
"Time/CNN Poll: Evolving Perspectives," *Time* (June 11, 2001): 46–47.
Tölölyan, Khachig. "Rethinking Diaspora(s): Stateless Power in the Transnational Movement," *Diaspora* 5.1 (1996): 3–35.
Traffic. Dir. Steven Soderbergh. Perf. Michael Douglas, Don Cheadle, Benicio del Toro. USA Films, 2000.
Tsing, Anna. "The Global Situation," *Cultural Anthropology* 15.3 (August 2000): 327–360.
Turner, Graeme. *British Cultural Studies: An Introduction.* New York: Routledge, 1990.

Wald, Priscilla. "Minefields and Meeting Grounds: Transnational Analyses and American Studies," *ALH* 110.1 (1998): 199–218.

Waters, Malcolm. *Globalization.* New York: Routledge, 1995.

"Welcome to Amexica," *Time* Special Issue. (June 11, 2001).

Werbner, Pnina and Tariq Modood, eds. *Debating Cultural Hybridity: Multi-Cultural Identities and the Politics of Anti-Racism.* London: Zed Books, 1997.

Williams, Fred O. "Closing the Currency Gap," *Buffalo News* (May 27, 2001): C-5 and C-7.

Wilson, Thomas M. and Hastings Donnan. *Border Identities: Nations and States at International Frontiers.* Cambridge: Cambridge University Press, 1998.

Young, Robert J. C. *Colonial Desire: Hybridity in Theory, Culture, and Race.* London: Routledge, 1995.

1. Border Theories

Border Shopping

American Studies and the Anti-Nation

Bryce Traister

> Then what about me
> what about the I
> confronting you on that border you are always trying to cross?
>
> —Margaret Atwood ("Backdrop" 51)

I began writing this essay one day after Seattle riot police "managed" the crowds protesting the opening of the 2000 World Trade Organization (WTO) Millennium talks. The WTO convened these talks ostensibly to review and renew the series of multilateral trade agreements that the WTO oversees, if not to simulate a liberal democratic public dialogue within the context of late capitalism. By the late fall of 1999, the WTO had, for many, come to symbolize globalization in its purest and most threatening form. It is a non-governmental organization (NGO) operating beyond the control of the nation-states (and their citizens) who are its signatories. Yet at the same time it retains control over and exerts control within those nation-states, and not just in the increasingly diverse realm of "trade," as the zealous police response to the conference disruption made alarmingly clear. While the WTO and many government officials insist that the organization, and globalization more generally, benefit the economic interests even of latte-sipping protesters (and seem to approve the use of teargas and pepper-spray in order to get that message across), the recent history of the WTO suggests that some of these fears are well founded.

In July 1997, the WTO appeals board issued a ruling that struck down a Canadian government policy of denying distribution rights to U.S.-owned magazines that failed to have a specified amount of "Canadian editorial content." Although "Canadian content" is a nebulous term referring to an even vaguer set of cultural practices, it denotes a discourse by or for Canadians about issues pertaining to Canadian "culture" in the widest sense of the term.[1] The U.S. government, arguing on behalf of U.S. corporations in the only global tribunal with enforcement powers, successfully "demonstrated" that the Canadian statute, although passed by majority vote of a democratically elected national government, violated the principles of fair trade enforced by the WTO and agreed to by a previous Canadian government. In short, the WTO ruled in favor of the United States and its corporations and against the current Canadian government in the name of ensuring, to quote from the WTO mission statement, "that trade flow as smoothly, predictably and freely as possible" across and through international borders.[2]

That the WTO, and globalist ideology in general, finds little meaning in local political measures and in the borders that define the extent of those measures is hardly unexpected. A more surprising phenomenon is that the anti-national and anti-territorial biases of corporate globalism have found re-articulation in much recent theorizing within the precinct of American studies. Take, for example, Paul Jay's recent discussion of American literary and cultural "border studies," where he "endors[es] the idea that our criticism can best be revitalized by paying more attention to the locations that are *between* or which *transgress* conventional national borders" (167). In her 1998 presidential address to the American Studies Association, Janice Radway similarly wondered "[i]f the notion of bounded national territory and a concomittant national identity deriving isomorphically from it are called into question, why perpetuate a specifically 'American' studies? Has enough work been done at this point to complicate and fracture the very idea of an 'American' nation, culture, and subject?" (16–17). Radway's answer (and Jay's for that matter) derive to some extent from reading important borderlands thinkers like Gloria Anzaldúa and José David Saldívar, as Carolyn Porter's synthetic critique of the developing post-national and postcolonial tendencies in American cultural studies makes clear. In her view, the "real work" of borderlands critics like Saldívar's is that it brings "'the school of Caliban' to the forefront of an American cultural studies that is thereby radically reconfigured as a field" (Porter 504). In place of a monolithic conception of a static national entity, border critics propose a concept of culture that is inherently fluid, unfixed, and heterogeneous. As I will show in the next section, such work urges us to imagine an American cultural studies model based not on territorial specificity so much as on discursive diversity; not on the nationhood of literature so much as on literature's deconstruction of the nation; and not on the identifiable culture(s) within clearly demarcated borders so much as on the borderlands cultures that, in resisting desires for homogeneous nationhood, expose the artificiality of all such nationalizing desires.[3]

The attack on national borders made within these left intellectual traditions, I will suggest, has an unlikely—one hesitates to say unwitting—ally in what we might call the weak nation desires of corporate globalism. Like globalist ideologues declaring the irrelevance of the nation-state and its borders and celebrating the beauties of transnational capital accumulation, the post-national agenda in border studies has dispensed with America's borders in order to pursue a cultural studies independent of the definitions, ideals, and, to be sure, the repressive apparatuses of the U.S. nation-state. At the same time, in the neoliberal discourse of global capitalism, the significance of the nation-state—as an imaginary construct soliciting obligation from the citizenry and as an assemblage of institutional realities that impact the lives of national subjects—has diminished significantly, even as the repressive capacities of a weakened nation-state have taken on other forms of discipline in the transnational globalist environment.[4] I argue that the importation of a border studies model into a continually revised Americanist field-imaginary has replicated some of the fundamental tenets of contemporary corporate globalism including its emphasis on consumerism. After a brief discussion of the assimilation of borderlands theory into Americanist critique, this essay will turn to a consideration of the Canada-U.S. border as a contested site of nationalist and globalist ideology before concluding by suggesting that some tendencies within contemporary Canadian nationalism provide a possible alternative to the erasure of meaningful national difference. Missing from the post-national attack on American studies that has grown out of Mexico-U.S. border studies is a sense of how a stable—indeed, fixed, unwavering, and meaningful—Canada-U.S. border may serve the useful purpose of containing the United States within the limits of its own boundaries and of forcing the expanding and increasingly corporate U.S. imperialism to stand in the light of recognition.

From Nation to Hemisphere

It is worth insisting at the outset on the difference between the Mexico-U.S. and the Canada-U.S. borders as lines of national identification. The Southwest and northern U.S. territorial borders present vastly different twentieth-century historical narratives: the former a contested history that has now become a unilaterally militarized struggle; the latter a history of economic exchange (and, in the nineteenth century, military conflict) which, although not without the usual struggles born of economic dominance and subordination, has proceeded more or less peacefully and within the "friendly" universe of modern and late capitalist social exchange-relations.[5] So while *la*

Frontera—the borderlands of the U.S. Southwest/northern Mexico and the site of much recent theorizing of a post-nationalist borderlands critique—solicits conceptualization of a more fluid exchange of identity across borders, the northern United States/southern Canada border presents a different set of problems to negotiate and articulate as a critical borderlands practice.

In short, some Canadians like the border just the way it is: permeable for the purpose of circulating goods, services, and persons, and fixed for the purpose of articulating national difference. In a recent meditation on the place of borders in contemporary Canadian thought, W. H. New sketches the delicate balancing act the 49th parallel demands of Canadian national identity: "Borders, as sites of contestation . . . neither require nor guarantee fixed differences, or inevitably commit to the erasure of difference . . . the presence of the United States right next to Canada almost constantly presents Canadians with socio-political options: some of which they adopt, some they resist, and some they . . . export" (New 27). Perhaps another way to put this is that, while the borderlands of the 49th parallel make possible cultural and economic networks of trans-national exchange for Canadians and Americans alike, the border itself tends to refract and distribute identity into nationally identified registers. This latter tendency may not, in fact, be altogether a bad thing. New wonders in relation to the nation/empire-building nostalgia of Canadian Studies and its desire for a cohesive set of national identity scripts: "Should we give away the state—which is a still working, if sometimes creaky, set of social agreements—because some people are impatient with it and other are ill-informed?" (17). I would add to this list of reasons for disliking the nation-state that of ongoing institutional marginalization and exclusion of those still deemed Other to the state.

As such, the liberal state has become something of a problem in post-national work in American Studies, which has declared the state and its borders to be in various ways illegitimate or perniciously "nationalist" in their tendencies. As I have argued elsewhere, in its anti-nationalism, much of this post-nationalist and postcolonial theory stages a straw argument about the "current" status of American literary and cultural critical practices (Traister 197–8). To some degree, this assertion of a totalizing and totalized American Ideal follows from the appropriation, by liberal-progressive academics, of the post-national and postcolonial critique opened up in work by third-world, Caribbean, and borderlands critics. In *Border Matters,* his follow-up to *The Dialectics of Our America,* José David Saldívar positions a specifically non Anglo-American critical field in relation to its imperializing other. Analyzing the cultures of the urban barrio and the Chicano garage bands of San José, California, Saldívar praises the borderlands as a space that is inherently resistant to being reduced to a "national tradition" (12), as conducive to a cultural subjectivity born of struggle and resistance rather than consensus and coherence

(14). "*Border Matters,*" he writes, "challenges this stable naturalized and hegemonic status" of the United States and its territorial stability. As a subdivision of Chicano/a cultural studies, border studies "offer . . . a deconstruction of the discourse of boundaries" (25), a reconceptualization not merely of the historical/material practices of the present-day U.S. Southwest, but of the very notion of international borders and the nation-states they frame.

What might be called an anti-national agenda has long flowered in borderlands academic and activist culture of the southwest United States. According to Chicana theorist and activist Gloria Anzaldúa, "[b]orders are set up to define the places that are safe and unsafe, to distinguish us from them . . . [the border] . . . is a vague and undetermined place created by the motional residue of an unnatural boundary. It is in a constant state of transition" (Anzaldúa 25). The relation between borderlands and nation becomes more than merely oppositional in nature. Within the space between, we learn about the constructed nature of life on either side of the divide as much as we encounter the suppressed or forgotten voices of *la frontera.* In his 1992 autobiography *The Other Side,* Ruben Martinez, for example, writes that he came to Tijuana, Mexico "thinking that perhaps by studying this city, I will learn something about Los Angeles" (83). What is exciting about the work of Anzaldúa and other Chicano/a and borderlands theorists is the rich critical praxis achieved in the explication of border theory from the Mexico-U.S. borderlands themselves. With respect, I would submit, however, that what "we" learn from this work is not merely the fluid nature of national identity or national claims upon identity, but also how to talk back to those claims, how to answer them, how to refashion them.[6] But this is a bit different from claiming that there really is or ought not to be any such national culture or set of cultures or dominant culture, trying to interpolate us as subjected citizens. Indeed, the border emerges as significant precisely to the degree that it registers nationally inflected scenes of cultural difference, even if its ultimate cultural and political work tends to collapse, blur or deconstruct such difference as merely perceptual, performative, imaginary, socially constructed, contingent, and/or just plain false.

What post-Americanist critics have derived from critique of the Mexico-U.S. border, then, is a *model for doing literary and cultural criticism of North American culture.* "By examining the contact zones of the Mexico-U.S. border," writes Saldívar in *Border Matters,* "we can begin to problematize the notion that the nation is 'naturally' there: These are spaces within which patronymic relationships take place" (14). Certainly, a point like this would be very familiar to scholars of North American indigenous culture, as evidenced in Donald Grinde's piece in this collection, which is to observe that the anti-imperialism animating much border and post-national critique informs the attack on the transcendent nation in fairly self-evident ways.

Moreover, the incorporation of a specifically U.S. Southwestern/northern Mexican materialist critical practice into a hemispheric model performs the clearly undesirable conceptual move of reifying three nations into one. The erasure of national difference undertaken in the assertion of such a borderlands model has remained, so far, an unspoken issue in the borderlands/postnationalist critique of the United States.

If from a different perspective—the one that notes that 60 percent of the population of Canada lives within 100 miles of the Canada-U.S. border—the entire nation of Canada may be regarded as a "borderland," the most appropriate question to ask might be: What kind of borderland is being discussed in post-nationalist American Studies? And whose borderland is it? In the next section of the essay, I consider the possible place of a Canada-U.S. border encounter—specifically border shopping—in the ongoing reconceptualization of the Americanist field imaginary. I will suggest that much of what appears in the narratives of border studies about the U.S. Southwest as a politically progressive deconstruction of U.S. police-state border ideology manifests itself as a globalist attack on the national integrity of Canada along the Canada-U.S. border.

Border Shopping I: Canadians Perform America

The Canada-U.S. borderlands are, among other things, a good place to shop.[7] The traditional structure of Canada-U.S. border shopping has been organized around the principle of choice and price, and historically the phenomenon of cross-border shopping has taken the form of Canadian citizens taking their consumer desires to U.S. stores boasting greater product choice and lower prices. Since the signing of the North American Free Trade Agreement and the simultaneous plunge of the value of the Canadian dollar relative to the U.S. greenback, the phenomenon of Canadians shopping America has been a less pronounced feature of public policy discussions north of the border; at the same time, the strength of the American dollar has solicited American border shopping in cosmopolitan Canadian urban centers like Toronto, Montreal, and Vancouver. Traditionally, while not the metropolitan centers typically associated with the American urban sublime, Bellingham, Buffalo, Detroit, Burlington, and Port Huron, have, however, continued to factor importantly in the Canadian imagination of what the United States is: A mecca of both nondurable goods like gasoline, milk, and poultry, and durable goods like clothing and entertainment items whose very diversity and cheapness might enable the average Canadian citizen to fulfill consumer desires in a way that the Canadian retail environment, with its hidden taxes and higher prices, could not.[8]

In the late 1980s, an increasingly unpopular Canadian government brought in a 7 percent Goods and Services Tax (known as "the GST"), a move which, although rendering visible a tax that had for years been invisible, intensified Canadian consumer desire for the U.S. shopping environment (Canadian Chamber of Commerce 7). What we might say is that the GST had the effect of aggravating, if not precisely anti-Canadian then at least anti-government sentiment amongst its citizens, and soliciting, within the matrix of Canadian national identity, the production of American identifications refracted through consumer desire. That is, the GST rendered visible what had been a dormant aspect of Canadian national identity: a desire for the apparently unlimited, unfettered, and untaxed choice enjoyed by the modern American individual (New Brunswick 11). In a sense, the GST could be said to have produced Americanized subjects within Canada, as the affiliations of citizenship within Canada and among its citizens became increasingly and visibly organized around a consumerist individualism identified with the U.S. and against Canada, whose very "Canadian" and expensive public institutions—universal healthcare, for example—were the driving (or shall we say sucking) forces behind the GST's necessary creation. The GST solicited American identifications in a way that disclosed the antagonism of border-crossing Canadians toward their own projects of Canadian civic nationalism. Both the NAFTA and the GST have together facilitated the kind of economic traversing of borders celebrated in globalist ideology, and in this regard the diminishment of border integrity—as barrier or point of cultural or national demarcation—has been underwritten by the economics of global exchange cultures that is based on producing a universal consumer subject.[9]

To be sure, Canadian national identity has always manifested itself in dialectical relation to its southern neighbor; Canadian Confederation, as it is taught in schools in Canada, resulted at least partly from fears at the conclusion of the American Civil War that a reunified United States would attempt to assert continental dominance. As Canadian educators and intellectuals have long been aware, the historical cultures of the U.S. have become "part of" the political, legal and historical imagination of Canadian subjectivity far more readily than have Canadian versions of the same achieved resonance south of the border.[10] But what is different about the Canadian (mis)identification with the United States today is that it increasingly takes the form of consumer rather than civic affiliation, as those political, legal, and even historical points of differentiation fade into the background only to play a supporting role in the staging of a North American homogeneity.

In late winter 2000, for example, the issue of Canada's nationalized medicine again became a popular topic in national and regional media, with many arguing that Canada must adopt a more "American" system of healthcare. With insurance companies offering preferential treatment for those who can afford it, an Americanized "two-tier" healthcare system would reclassify healthcare as one among many consumer options; formerly a social institution associated with Canadian civic nationalism and actual Canadian bodies, medical healthcare would become just another mecca of "choice," another occasion for commercial persiflage, and yet another site for the unfettered assertion of monied rather than national agency. Indeed, healthcare has already become a more visible cross-border phenomenon in Canada, as government hospital closures, funding cutbacks, and "restructuring" have led to Ontario cancer patients having to travel to Buffalo, New York, for radiation treatment. Some would say that as the single most important civic institution of the nation, the Canadian health system is increasingly under pressure to surrender its control of nearly every aspect of healthcare to the corporate profit motive. As healthcare becomes yet another occasion for "border shopping," the civic institution widely cited as symbolizing the Canadian nation-state potentially becomes another irrelevant institution in a post-national globalist order.

The fantasy of border shopping, at least in the eyes of Canadian consumers trundling back and forth across the border with new underpants worn under new blue jeans worn under bulky sweaters worn under new parkas, not only stages a critique of Canada from without, but more dramatically affirms America from within.[11] The permeability of the border, as manifested in the cross-border shopping phenomenon, can be said to make a case for the constructed status of the "borderlands" and the nation it nominally defines. "Citizenship" here becomes a performative category of identity formation, as the national citizen manifests itself within and as discourses of consumerism ostensibly independent of the claims of national and nationalist subjection. The consuming subject subsists in relation to the a-national goods and services gathered during shopping excursions in America, and thus resembles those Benetton ads in which ethnically diverse and always smiling children promote an inveterately North American palette of clothing style and color as gloriously innocent global diversity.

To some degree, the performative staging of citizenship enacted by the Canadian-American consumer ironically manifests a conception of identity commensurate with critiques of national literary and cultural identity inherent in postcolonial and post-national borderlands critique. Canadian border shopping, after all, assumes a certain economic and class status insofar as the would-be Canadian consumer of American retail possesses both the financial means and the consumer savvy to pursue shopping opportunities in the U.S. borderlands. More speculatively, we might say that the presumably affluent Canadian border-crossing shopper of North American late capitalism constructs the hinterlands of the Canada-U.S. border as an effect of a certain kind of economic and cultural privilege that essentially ignores national affiliation, while it articulates a rhetoric of identity commensurate with American national mythology.

In the realm of the academic theorization of border fluidity, it is worth noting at this point that the border-crossing desires of a post-national American Studies discloses a hitherto undertheorized privilege of U.S. academia: that of imagining, in abstraction at least, the borderless world as a series of cultural spaces in which a post-Americanist critical imaginary might articulate its own politically necessary vision of identity. According to John Muthyala, "Geographically and historically, the border regions that lend a special focus to contemporary theorizations of border experience in North America compose the U.S. Southwest, those parts of Northern Mexico which the United States annexed with the signing of the Treaty of Guadalupe Hidalgo in 1848, namely California, New Mexico, Arizona, and parts of Nevada, Colorado, and Utah" (111). What interests me here is not the narrowness of the claim—which is conceptually troubling in itself—but the rhetoric of its deployment: the "special" status accorded to a certain, regionally defined account of history, narrative and experience, whose re-configuration not only in, but, more crucially, *as American studies,* produces a discipline "truly global in scope" (114). According to John Carlos Rowe, "America's conceptual and geographic boundaries" become "fluid, contested, and historically changing" (15). Although we're still talking about the United States of America in all of its "special" reconfigurations, it now appears that America, under the auspices of these new regional/area studies re-mappings, would become more or less simultaneous with the world itself. As Carolyn Porter worried a few years ago, the post-national or meta-territorial theorization of American cultural identity might solicit in academic terms precisely the imperializing sorts of claims such critiques are intended to resist (Porter 523). And at what point in Lawrence Buell's speculative promotion of America as postcolonial critical project does one begin to suspect that well-meaning anti-imperialism comes close to refashioning postcolonial critique as "cutting edge" Americanist cultural theory?[12]

Even though it appears to be a perfect manifestation of post-nationalist theorizing, Canadian cross-border shopping ultimately entails acts that are imitative or iterative of various American national or nationalist scripts. The plethora of abundance and diversity encountered in the United States, historically the scene of frontier mythology, agrarian self-sufficiency, and democratic possibility, emerges as terrestrial paradise of consumer splendor and as evidence of America's successful destiny. The irony of American studies critics positing, commemorating, or desiring the expansion of their academic subject beyond that of "America" emerges when we consider the degree to which Canadian consumers bring America back to itself in the form of a cross-border shopping subject. This subject reaffirms the promise of American bounty and democratic individualism with every gallon of gasoline purchased under relaxed, environmentally damaging, and corporate-approved tax law. That consumer, I would say, is first and foremost construed as an "American" subject, whether or not she's wearing an American flag or a Canadian maple leaf on her new denim jacket. The "global citizen" of a post-nationalist imaginary similarly conjures a subject born of American-style privilege. Emptying the border of a content that is distinct, or ostensibly distinct, from consumer culture would leave the Canadian border zone bereft of its own national, nationalist, and nationalizing content. Under the "area studies" conceptualizations of this new post-national critical imaginary, Canada would become what many of its citizens fear it already has: a zone part of and therefore subordinate to the United States.

To put it bluntly, Canadian nationalist endeavors, often undertaken by government and government agencies, attempt to protect and promote the interests of its citizens as part of its ordinary operation as a duly elected parliamentary political entity. Take, for example, the issue of residential import/export business between Canada and the United States. In order to generate revenue and encourage Canadian residents to "buy locally," the Canadian government applies the GST and other applicable tariffs to goods purchased in the U.S. and shipped into Canada. The shipping company that delivers packages collects the tariffs from the receiver, and collects, in addition to the tariff, a fee for "brokering" the shipment. Different shipping companies charge different fees for collecting the GST and any other applicable tax on behalf of the Canadian government. Canada Post, an independent corporation subsidized by the Canadian national government, charges a flat fee of $5; UPS charges a fee based on the declared value of the shipment. The UPS fees are, in most cases, significantly higher than those charged by Canada Post. In other words, UPS believes itself entitled to make a profit by brokering goods through the Canadian tariff system.[13] Canada Post and the government that partially subsidizes its operations through—you guessed it—the taxes it collects from its citizens and residents receiving international shipments, would have a hard time explaining to its outraged citizens and residents why, in addition to the taxes payable on the import, they must pay an additional amount for having the taxes collected in the first place. The point here is not merely the obvious one—that nations and citizens are answerable to each other in ways that corporations and consumers are not—but that the contract of answerability enjoyed by subjects of the nation, while fraught, contested, and often unequally distributed across the citizenry, makes a case for retaining (rather than repudiating) the nation and national identity as a significant feature of cultural critique and inquiry.

Canada Post operates the way it does not merely because it would have a much more difficult time getting away with what some might call highway robbery, but because the attitude informing its "business" practices is part of a wider, and indeed national, network of attitudes in Canada about the nature and practice of good government, and the sort of beliefs about self and society such national(ist) practices promote and reflect.

Border Shopping II: Americans Do Canada

Americans also "shop" Canada, and they do so, in part, by setting up retail outfits inside its borders. Technically "owned" by Canadian holding companies, thus rendering the obvious lie of "Canadian owned and operated" into stylized rather than illegal ad copy, corporations like Wal-Mart have been known to festoon their stores with Canadian flags, "specialize" in product-retailing tailored to Canadian "cultural" realities like snow, hockey, and cheap candy, and loudly insist that they become part of the Canadian civic landscape in which they erect their enormous parking lots and the stores in which all employees—"true" Canadians every last non-unionized one of them—are required to smile. ("At least they're shopping in Canada," I heard somebody say once, thereby buying into the Wal-Mart claim that they bring jobs to Canadian communities.) These days, what ultimately matters to first-world consumers is retail, and from border shopping, to Wal-Mart growing, to Starbucks brewing, the desire for national affiliation and difference has come to be associated with the idea that trade protectionism of any sort, and the desire for national self-interest such measures have in the past tried to satisfy, have become an expense "the people" can simply no longer afford. The discourse of national identity in Canada that organizes itself around a public welfare state thus emerges as the foe to be slain by the new universal consumer, a figure whose attitude toward the nation-state most closely resembles a socio-political libertarianism first articulated in the early essays of Emerson.

In a chilling 1993 essay, "'Globalization' and the University," Masao Miyoshi powerfully articulates this phenomenon by observing that "[i]f there is an identifiable style in TNC [transnational corporate] culture, it is 'universal' consumerism that spreads beyond the boundaries of the first world into the second and third, providing they have enough money left over to spend" (259). Americans shopping in Canada are no less American for doing so; indeed, they are even more inveterately and even mythologically American in their exertion of spending and cultural power, in their exercise of personal freedom and autonomy to render their interior state visible as exterior estate. This is to say that the weak-nation theory advanced by Americanist theory imagines the annihilation of the nation and its simultaneous public domains for a post-national non-consumer based citizenship, but does not account for those forms and mythologies that inhere within the spaces colonized by the new globalized consumer.

In any "expanded" or "re-drawn" map of American Studies in which Canadian culture plays a role, the "Canadian-ness" of any such investigation will mean vastly different things, depending on the conceptual framework governing the analysis. In a post-nationalist "Borderlands" study of Margaret Atwood's *A Handmaid's Tale,* for example, what might be the meaning or significance of the evangelical Christian theocratic "republic" of Gilead? In the view of many Canadian literary and cultural critics (including the novel's author), the fictional Gilead's possible resemblance to certain literalizations of specifically American historical and cultural streams of experience is no mere coincidence, and is designed to probe precisely the stain of "exceptionalism" many contemporary Americanists would have us wash out of the fabric of American studies. Is the *Handmaid's Tale* a "Canadian" or "American" novel of the North American borderlands? In the conceptual apparatus of borderlands studies, such a question would be irrelevant, if not obnoxious in its prescription that some cultural artefact be of one nation or another. While "binaristic" in its implications, however, the question remains pertinent in terms much contemporary borderlands critique would have a difficult time answering.

In 1993, the universal ethos of citizen-based consumerism presented for Miyoshi "the theoretical possibility that regional cultures everywhere may be obliterated before long" (259), a possibility that has, at least here in Canada, become a chilling reality. One such culture, whose implications are both regional and national in scope, is the distinctive legal culture of Canada, whose difference from the U.S. legal system became clearly manifest in the early 1990s during the trials of Canadian serial killer Paul Bernardo and his accomplice and soon to be ex-wife, Karla Homolka. In the wider media event that the brutal sex murders of teenagers Leslie Mahaffey, Kristin French, and Tammy Homolka (Karla's younger sister) came to generate, the right to consume media stories of the killings manifested themselves as the rights of democratic subjectivity. In two celebrated moments of American corporate imperialism, both the *New York Times* and the *Washington Post* editorialized that the Canadian ban on publishing the details of the case violated the principles of free speech and established a "tyranny" of government censorship incommensurate with the "universal" rights of the democratic nation-state. With a lack of self-reflection and sense of righteous entitlement that the world has come to associate first and foremost with American identity par excellence, American corporate media institutions unwittingly called attention to the America hidden within values cited for their paradigmatic universalism.[14] Democracy, as American TNCs and the WTO have become fond of reminding the world, is always good so long as it resembles the sort practiced in America.[15]

Canadian media conglomerates and writers, also inflamed with the desire for the increased circulation and fame the sordid tale would generate, joined the American corporate national agenda by asserting the American ideology of a "free press" as their own. Not for the last time, to be sure, American and Canadian corporate institutions asserted the right to consume within a rhetoric of the rights associated with specifically American legal and political institutions. The assertion of American legal and political institutions as universal is thus not merely coextensive with the right to consume, but more dramatically disseminates "Americanness" as the "true" interior of other national institutions and subjectivities. In the end, following Tocqueville, we're all just Americans looking for bargains, scandal, and, to be sure, lower taxes.

The Anti-Americanist Agenda

The ideological and consumer desire circulated during the Bernardo-Homolka affair finds more "literary" expression in the split-run magazine contest staged between the Canadian government and American corporations wanting to sell magazines boasting Canadian advertisements and a dearth of Canadian editorial content, with which I began this essay. At stake in the struggle was the marketability of "Canadian cultural content," including everything from art and literary practices to Canadian local, regional, and national politics. In one sense, to be sure, the struggle revealed the "constructedness" of Canadian culture, with its dependence on government subsidy and granting agencies, and the difficulties characteristically encountered in the attempted articulation of "Canadian identity" premised on some category other than "not-American." It would be foolish as well to imagine the Canadian government's interest in fostering Canadian cultural expression as the purely disinterested act of fostering civic harmony, especially since the immediate interests defended in the WTO magazine dispute were first and foremost corporate.

The Canadian-American border, as a site of national differentiation, has become yet another casualty in the global march into the universalist ethic of *Sports Illustrated,* a text whose sense of cultural diversity is best reflected in its U.S. advertisements featuring multi-millionaire African American athletic superstars wearing shoes assembled in Indonesian sweat-shops. In its ruling against the Canadian position, the WTO didn't merely demonstrate the irrelevance of the border as a place of meaningful demarcation of identity and practice; it called attention to the border's artificiality as a space that refracts and separates national identities from one another. According to Dennis Browne's review of the split-run magazine case, the WTO arbitration panel "considered the examples of split-runs published in Canada by *Time* and *Newsweek* and concluded that a news magazine is a news magazine irrespective of its source" (Browne 22). The magazine's commodity status overrides its status as a nationalistically inflected cultural artefact; indeed, it emerges as a moment in globalization when, according to Fredric Jameson, "we begin to fill in the empty signifier with visions of financial transfers and investments all over the world" (56).

Whether or not one accepts the unlikely premise that the Canadian government had cultural rather than corporate interests at heart in its fight with the United States/WTO, we can all agree that the magazine contest referred to both financial and cultural exchange. The postal subsidy given to Canadian magazines (one of the central bones of contention for U.S. media giants like Time-Warner), was designed to make the circulation of Canadian news magazines like *Saturday Night* and *MacLean's* less financially crippling. Within the pages of such magazines we find, no matter how nebulous a concept to define, "Canadian content" in its most current and arguably generically diverse form. The WTO ruling against the postal subsidy, which underwrote the comparatively low and expensive circulation of such magazines, effectively disrupts the circulation of Canadian culture within its own borders, while opening the door further for the circulation of U.S.-centered magazines like *Time* and *Newsweek* within Canada.[16] It is thus not merely the case that globalism has contributed to the American purchase of Canada in the form of big business retail; globalism has also forced local civic and community institutions to abandon measures and policies that would promote national or nationalist endeavors that might run counter to the weak-border/weak-nation impulses of globalist ideology.

The frank nationalism of the Canadian position, articulated as government support of a corporate or civic agenda, has characterized much of the discussion north of the border, while to the south post-Americanist academicians have denounced the claims of nation on the disciplinary imagination. In this sense, the magazine conflict materialized Janice Radway's warning that "[i]f intellectual practice in the field does not examine the ways in which the construction of a national subject works to the economic and political advantage of some and precisely against the interests of others, then American studies runs the risk of functioning as just another technology of nationalism, a way of ritually repeating the claims of nationalism by assuming it as an autonomous given inevitably worthy of scholarly study" (Radway 12). With Radway, other scholars have all too often premised critiques of what Donald Pease has so usefully called the "Americanist field-imaginary" on the idea that nations and nationalisms, national subjects and subjects of/to nationalism, are for all intents and purposes interchangeable, and, more to the point, just altogether bad.[17] Radway's suggestion that we radically reformulate the field and practice of American studies independently of geographical boundaries and all other markers of static, monological fixedness potentially threatens to render the intellectual work of these fields into the handmaiden of globalism's anti-national imaginary.

Let me be clear here. The stated and suspected political commitments of academic theorists of a post-nationalist borderlands sublime in no way resemble or accord with those of committed global capitalist apologists like Federal Reserve Chairman Alan Greenspan or soon to be former Canadian anglophile robber-baron Conrad Black. This scholarship is not in any deliberate way complicit with corporate globalism's attack on participatory democracy and the public sphere. At the same time, we must recognize that what academic and activist critics celebrate in the borderlands of the southwest United States and propose as the "new" model of doing Americanist scholarship, global capitalists laud in southwest Ontario (a province in Canada): the emergence of a nation or nationalism whose claims on its citizens, subjects, or consumers have become so attenuated or "constructed" that the nation plays second fiddle to big American business. "Ontario is now open for business," the government of right-wing ideologue and corporate apologist Mike Harris announced, as his government gutted government ministries of the environment, privatized publicly run and regulated utilities, and rhetorically savaged the idea of the liberal state and its employees.[18]

From within the left-progressive domain of contemporary Americanist cultural theory, we detect the strains of a language commemorating as elegy not merely the ideological construction of American literature, culture, and history, but also celebrating the very passing of the nation-state and its cultural markers. American border shopping in Canada, and the laissez-faire multi-lateral trading environment that makes it so rewarding a shopping experience for well-heeled Canadians, could be said to advance a "weak nation" future as consumer desires, identification, and practices come to replace or at least suspend national or communal affiliations. The ethnoanthropological critique of the southwestern United States, while advancing entirely different politics than those of the WTO, similarly exhibits a weak nation desire for a nation-state whose claims on its citizens would be, at best, contingent and secondary to other affiliations (like ethnicity) by which meaningful civic and political coalition might be established. What the example of the magazine contest between Canada and Time-Warner reveals, however, is the need for a renewed encounter with rather than relinquishment of claims made by, on behalf of, and against the liberal nation-state. These claims are not merely made by self-evidently, perniciously nationalist discourse on its interpolated subjects, but solicited by actual subjects of those national discourses in their material and cultural practices that perform nation and national identity in the doing and the telling.

The obvious parallel to draw, of course, is between the WTO's evacuation of the magazine text as a site for staging and circulating Canadian national identity within its own borders, and the proposed evacuation of United States literary or cultural texts as scenes of identifiable American formations. Indeed, the anti-nationalist's antagonistic relation to American exceptionalism, like the WTO's denial of national(ist) efforts at cultural self-definition, has issued something like a declaration against the idea that *any* nation might be exceptional: That is, constituted in uniquely self-referential or culturally and nationally specific terms.[19] For many others, whether engaged in the intellectual labor of Canadian studies or trying to sell a brand of beer called Molson Canadian, locating the Canadian national within the North American—that is to say, U.S.—cultural imaginary is a vital undertaking, and one implicitly denied in much borderlands theory.[20] I would suggest that the deconstruction of the idea and borders of the U.S. nation-state and, by definition, of the nations arrayed on its territorial borders, exhibits the very *privilege* of Americanist theoreticians in relation to other North Americans, for whom what John Carlos Rowe has called the "national cultural boundaries" (12) of the nations of North America can mean very different things. Unlike nearly every other theorization of American border/area studies published in the last five years, Rowe manages to name Canada as one of the "different nationalities, cultures, and languages of the Western hemisphere" (13). But even his North American capaciousness is not without conceptual difficulty, as he goes on to propose that "[t]his new interest in border studies should include investigations of how the many different Americ*as* and Canad*a* have historically influenced and interpreted each other" (13–14; emphasis mine). As an officially bilingual country established in the historical, political, and ideological fulcrum of precisely the multicultural, multilingual, multiethnic, multi-religious, and multi-regional borderlands of the United States current theories of American studies are encouraging us to include in a re-mapping of the discipline, Canada remains the reified singularity of an inveterately American perspective. While the difficulties encountered in imagining a Canadian national identity or the identities inhering within the border shopping zones of the Canada-U.S. border may be formidable to say the least, they nevertheless hold the promise of issuing an understanding of national identity that is distinct from the emptied versions of the nation urged by post-nationalist and globalist ideologies alike. In our laudable commitment to anti-hegemonic historiographical and theoretical scholarship, we have lost our ability to understand the liberal nation-state as a positive and still intriguing contributor rather than impediment to meaningful and even politically progressive identity. We also, in making vague pronouncements about considering "the Americas" next to monological Canada and enjoining others to embark on such scholarly voyaging, risk becoming the sort of imperializing tourist who, armed with the best intentions, still ends up condescending to its disciplinary or national other—in all essential points wondering how this or that country is or is not like our current critical version of America. Some would say that the sort of conceptual slip Rowe makes in his essay is exactly what the new scholarship of the borderlands will minimize by problematizing the "either/or" distinction implied by "America" and "Canada." I would reply that such a slip indicates that national difference remains a crucial register not only of culture and identity critique, but also of disciplinary relevance. Simply wanting to move the grasp of American Studies scholarship beyond U.S. borders because its currently "embordered" status represents a form of thinking we distrust fails to account for the persistent realities of meaningful cultural difference that national borders entail.

Historically considered, the weak nation desire of globalism has proceeded simultaneously with a rise in often alarming forms of tribal, ethnic, and balkanized nationalism. This apparently contradictory movement suggests to me that national and nationalist affiliations solicited and enacted in such contexts need to be rendered visible rather than only deconstructed into a post-national borderlands sublime. To be sure, the post-nationalist critique of American literary and cultural "homogeneity" proceeds from deeply held political and intellectual convictions that the idea of *one* Americanist cultural-imaginary advances the misogynist, xenophobic, homophobic sorts of nationalist ideology that animate its idealizing and thus exclusionary tendencies. Given the overdetermined forms of nationalist rhetoric voiced in so many arenas of American public life, one can hardly find fault with the motivation of the post-nationalist critique I have been describing here, particularly those voiced from within the marginalized borderlands/*frontera* cultures themselves. At the same time, however, it is worth pointing out that the formation of an American literary-cultural field-imaginary serves heuristic and conceptual purposes that are not simply co-extensive with banner-waving nationalism and other undesirable hegemonic projects. Crucially, a multi-faceted Americanist imaginary renders the boundaries of the United States, in both imaginary and geo-spatial terms, tangible and visible.

Perhaps the quiet, institutionally driven, and tax-elevated nationalism of Canada might serve as a useful rejoinder to the thorough abandonment of a public sphere in the United States implicitly urged by post-national cultural critics. Many of the political desires voiced on the intellectual left—universal healthcare, subsidized daycare, well-funded public education at all levels—exist at this time in most Canadian provinces, albeit in increasingly attenuated forms as the globalist attack on the civic ideal has met local support in the form of laissez-faire politicians genuflecting before the altar of the big-business elite. Notwithstanding the current situation, the idea of the Canadian nation-state as institutional protectorate has fostered the production of cultural projects purposively if confusingly committed to the articulation of meaningful Canadian national identity models. In the Canadian borderlands we find a way to keep the fluidity of "America" from achieving flashflood status.

Notes

1. The history of "Canadian content" ("cancon") is a long and complex one; like the idea of "culture," "cancon" has come to encompass a broad range of discursive practices including magazines, books, film, television, and now the Internet. As a legal mechanism, it dates to radio broadcast regulations of the early 1970s that governed the content of music and editorial content that Canada-based radio stations could broadcast. For a succinct recent account of the travails of "cancon" in Canadian popular culture, see Pevere and Dymond 168. Thanks to Peter Bailey for pointing me to this resource.
2. The citation is from the current mission-statement to be found on the WTO homepage (http://www.wto.org). The magazine dispute finally died when, in the fall of 1998, the Canadian government withdrew a new measure intended to replace the one declared out of order by the WTO.
3. In his trenchant analysis of the function of globalist ideology in contemporary American studies debate, Paul Giles remarks that "movements over the last decade to reconstitute the American literary canon can be seen as commensurate with the ideological subversion of nationalist paradigms by way of postcolonial discourses concerned with the expression of differential power equations" (Giles 525).
4. One thinks, for example, of the widely under-reported incidences of violent state military interventions in developing world labor disputes between workers and TNCs like Nike.
5. To be sure, the history of Canada-U.S. relations hasn't been conflict free, from British and French Canada's roles in the French-Indian war, to British Canada's role in the War of 1812, the underground railroad, and so forth. Still more recently, U.S. anti-immigration ideology has disclosed its own racism, as politicians from southwestern states have pointed out that the militarization of the Mexico-U.S. border stands in stark and unfair contrast to the most famous undefended border in the world. Canada-U.S. border relations have more recently turned sour over concerns about Canada's perceived lack of commitment to anti-terrorist and anti-drug activities. Apparently, the Canada-U.S. border more properly belongs to the United States of America, at least in the rhetoric of a recent congressional sub-committee investigating the "lenient" immigration policies of Canada.
6. For a reading of Martínez that focuses on the "cultural contestation, displacement, and reconversion" of the borderlands, see Saldívar's *Border Matters* 141–144.
7. I am not addressing in this essay the issue of aboriginal/first nations territorial border crossing, whose longitudinal tribal affiliations have historically transgressed and been divided by the Canada-U.S. border. A less polemical

version of this discussion of Canada-U.S. border relations might focus on how Native American/Canadian First Nation groups assert claims of sovereignty and national self-definition over and against the international border's putative separation of native community. Mohawk reserves straddling the international border in southeast Ontario and southern Quebec, for example, have exercised their sovereign right to unfettered trade to "move" tobacco products from the United States into Canada without paying or charging taxes. For a discussion of this issue, see "On the Fringes" and Donald Grinde's essay in this volume.

8. On taxation's contribution to the phenomenon, see, for example, New Brunswick 1992; Canadian Chamber of Commerce 1993.
9. In the early moments of 2000, one of the cultural conversations that dominated Canadian public forums revolved around the question of the "brain-drain": the widespread perception and/or reality of large numbers of Canadian citizens moving to the United States to "escape" the Canadian tax-burden and to revel in the capitalist splendors of the United States. Such discussion was invariably ideological at its core and ultimately organized around a desired shedding or acquisition of nationalist affiliation.
10. Canadian academic culture, particularly literary-critical debate over the nature of Canadian canonicity and cultural representativity, has also developed in the shadow of and through interaction with U.S. academic power. For a discussion of the U.S. presence in contemporary Canadian literary criticism, see Frank Davey, *Canadian Literary Power*, 45–78.
11. The image I'm conjuring here, of course, is a picture of cross-border tax evasion, another sort of "American" ideology that holds that evading taxes is nearly as patriotic a gesture as not having any tax burden at all.
12. I have in mind as well Buell's speculative attempt to resist what he calls "Americanist centripetalism" (415) by articulating American literature as postcolonial practice.
13. UPS charges what it calls an "entry prep" fee in addition to a percentage of the tariff/taxes it collects on behalf of Customs Canada (a government agency). On a shipment with a declared value of $50, UPS collects $3.75 in taxes payable to the Canadian government, $4.50 for collecting and paying the $3.75 in taxes payable, and $16.75 as an "entry prep" fee. Of course, the receiver of such a shipment has no say in the matter, and must pay these fees before receiving the package.
14. For this discussion of the media response to the murders, I am indebted to Frank Davey's *Karla's Web* and to our many conversations about the tragedy. It is worth quoting Davey's reading of the Bowering poem as it relates to some of the wider identity issues I've raised here: "Canadians are at war with the U.S. but cannot win. They surrender, hoping this will be a winning strategy. If they embrace the U.S. sufficiently, maybe it will 'get them off their backs' and allow them some independence. Maybe it will even come out 'in the light' where Canadians can finally see and learn to distinguish what is American and what is American" (212).

15. So when Canada greeted protesters of and participants in the 2001 Free Trade Area of the Americas (FTAA) Summit in Quebec City with a hastily constructed barricade around the conference site, suddenly American political and corporate leaders were impressed with Canada's ability to keep a lid on things.
16. Since the ruling, *Saturday Night Magazine* has become a dependent fifedom of the *National Post,* a national newspaper owned by mogul Conrad Black.
17. On this point see my "Risking Nationalism." Our oversimplified equation of nation and nationalism, although not directly attributable to the work of Benedict Anderson's *Imagined Communities,* certainly follows from a reading of his articulation of the nation's origin in imaginative and, hence, as many of his readers have found, "socially constructed" work.
18. When poisoned well-water claimed the lives of seven residents of Walkerton, Ontario, critics quickly identified the massive deregulation of state supervised water testing brought in with the Mike Harris government as the chief culprit in the tragedy. Either ignorant of or inspired by the year 2001 power crisis in California (the Ontario Tories would prefer to keep its citizenry in a state of darkness, after all), the Ontario government was still planning in 2001 to deregulate the province's state-owned power utilities, citing the inefficient and "unnatural" way that state-owned businesses run themselves.
19. I have in mind particularly the historiographical theory of Paul Gilroy, whose book *The Black Atlantic* can be seen, according to Paul Giles, as a fruitful result of the interchange between American studies and British cultural studies (Giles 540). Gilroy's consequent significance to later theorizations of the Americanist field-imaginary can be found, for example, in Janice Radway's claim that "the very notion of the U.S. nation and the very conception of American nationalism must now be understood as relational concepts" (Radway 29).
20. Such intellectual work would include that of Davey and New, among many others in Canadian literary studies. The Molson beer commercial features a young "twenty-something" self-identified Canadian male "talking back" to stereotypical conceptions of Canada and Canadians: "I am not a lumberjack or a fur trader, and I don't live in an igloo or eat blubber or own a dogsled, and I don't know Jimmy, Sally, or Susie from Canada, although I am certain they are really, really nice. I have a Prime Minister, not a President. I speak English and French, not American. And I pronounce it "about" . . . not "a-boot." Corporate sponsored articulations of "Canadian-ness" are not alone in stressing that Anglo-Canadians define themselves largely in negative relation to the United States.

Works Cited

Anderson, Benedict. *Imagined Communities.* London: Verso, 1983.

Anzaldúa, Gloria. *Borderlands/La Frontera: The New Mestiza.* San Francisco, CA: Aunt Lute Books, 1987.

Atwood, Margaret. "Backdrop Addresses Cowboy," *The Animals in that Country.* Boston, MA: Little Brown, 1968.
———. *The Handmaid's Tale.* New York: Ballatine Books, 1987.
Browne, Dennis. "Canada's Culture/Trade Quandary and the Magazine Case," *Canadian Parliamentary Review* 21.3 (Autumn 1998): 19–24.
Buell, Lawrence. "American Literary Emergence as a Postcolonial Phenomenon," *American Literary History* 4.2 (Fall 1992): 411–442.
Canadian Chamber of Commerce. "The Cross Border Shopping Issue: A Report," 1993.
Davey, Frank. *Canadian Literary Power.* Edmonton, Alberta: New West, 1994.
———. *Karla's Web: a Cultural Investigation of the Mahaffy-French Murders.* Toronto: Penguin, 1994.
Giles, Paul. "Virtual Americas: The Internationalization of American Studies and the Ideology of Exchange," *American Quarterly* 50 (1998): 523–547.
Gilroy, Paul. *The Black Atlantic: Modernity and Double Consciousness.* Cambridge, MA: Harvard University Press, 1993.
Jameson, Fredric. "Globalization as a Philosophical Issue," *The Cultures of Globalization.* Eds. Fredric Jameson and Masao Miyoshi. Durham, NC: Duke University Press, 1997: 54–77.
Jay, Gregory. "The End of 'American' Literature: Toward a Multicultural Practice," *The Canon in the Classroom: The Pedagogical Implications of Canon Revision in American Literature.* Ed. John Alberti. New York: Garland, 1995.
Jay, Paul. "The Myth of 'America' and the Politics of Location: Modernity, Border Studies and the Literatures of the Americas," *Arizona Quarterly* 54 (1998): 165–192.
Martínez, Ruben. *The Other Side.* London: Verso, 1992.
Miyoshi, Masao. "'Globalization,' Culture, and the University," *The Cultures of Globalization.* Eds. Fredric Jameson and Masao Miyoshi. Durham, NC: Duke University Press, 1997: 247–270.
Muthyala, John. "Reworlding America: the Globalization of American Studies," *Cultural Critique* 47 (Spring 2001): 91–119.
New, W. H. *Borderlands: How We Talk about Canada.* Vancouver: UBC Press, 1998.
New Brunswick Ministry of Economic Development and Tourism. "A Discussion Paper on Cross Border Shopping," January 1992.
Pease, Donald, ed. *National Identitites and Post-Americanist Narratives.* Durham, NC: Duke University Press, 1994.
Pevere, Geoff and Gregg Dymond. *Mondo Canuck: A Canadian Pop Culture Odyssey.* Toronto: Prentice Hall, 1996.
Porter, Carolyn. "What We Know We Don't Know: Remapping American Studies," *American Literary History* 6 (Fall 1994): 467–526.
Radway, Janice. "What's in a Name?" *American Quarterly* 51 (1999): 1–32.
Rowe, John Carlos. "Post-Nationalism, Globalism, and the New American Studies," *Cultural Critique* 40 (Fall 1998): 11–28.
Saldívar, José David. *The Dialectics of Our America: Genealogy, Cultural Critique, and Literary History.* Durham, NC: Duke University Press, 1991.

———. *Border Matters: Remapping American Cultural Studies.* Berkeley, CA: University of California Press, 1997.

Taylor, Rupert. "On the Fringes," *Canada and the World Backgrounder* 61 (1995): 26–27.

Traister, Bryce. "Risking Nationalism: NAFTA and the Limits of the New American Studies," *Canadian Review of American Studies* 27 (1997): 191–204.

Telling the Difference between the Border and the Borderlands

Materiality and Theoretical Practice

Manuel Luis Martinez

The international border between the United States and Mexico has served many purposes in its history. Its role in defining the American nation and its economy are but two of its functions. In the postwar period alone the borderlands have been figured as a magical space, a transformative site, the birthplace of *la nueva mestiza* (the Chicana/o subject), the *bracero* (guest worker), and the "new American hero of the western night." It is in crossing into Mexico near the end of the quintessential postwar American novel, *On the Road,* that Jack Kerouac's "countercultural" figures Sal Paradise and Dean Moriarty finally become archetypal Americans as they muse that the torch of martial spirited American individualism has been passed down to an increasingly anti-communalist postwar generation. Indeed, the torch has been passed by the American soldiers of the Mexican War who "cutting across with cannon" have provided the new generation of American adventurers with a road that "goes all the way to Panama," a road that "eventually leads to the whole world" (230). As the variety of postwar narratives about the border show, disconnected from its material location and its history of repression, exploitation and displacement, the borderlands can as easily function as the site of a New American Frontier as the site of the New Chicano Homeland, Aztlán.

The efforts of criticism on the borderlands to disrupt and resist master narratives of American nationalism have transformed the U.S.-Mexico borderlands into the birthplace of hybrid subjects, such as the Mexican migrant, the undocumented, and the marginal ethno-racial subject. This move toward a poststructural/postcolonial borderlands criticism as the most important and central set of practices and methodologies in contemporary Chicana/o studies has transformed the site of the border into a symbolic place by locating a purely discursive form of "opposition" in the U.S.-Mexico borderlands. The most damaging aspect of this so-called post-national work, however, is its dismissal of the importance of "place" and "citizenship" in a nationally defined entity that permanently denies especially the (undocumented) migrant arrival.

In this essay, I want to examine border narratives of such migrant workers, undocumented and displaced. They show us that migrants seek "to arrive" and to deploy their full civil rights within a responsive public sphere in a way that renews the idea of *Americanismo.* One of the most compelling and useful paradigms created by the Mexican-American generation of the 1930s and 1940s, *Americanismo* expressed the migrant's desire for "arrival" in what writer/activist Ernesto Galarza terms "*el pueblo libre*" (a free people/a free town) in his autobiographical novel *Barrio Boy.* Here, the term defines the imagined space where the migrant can live freely while establishing not only a *sense* of stability, but an actual *place* of stability. Rather than expand the liminal "borderlands," I argue that we should be focusing on the exploitation and constraint of the border itself, which has acted historically as the literal and figurative marker of nation-state exclusion, and as a literal representation of the denial of citizenship and its rights.

The purpose of my intervention is to point out ways in which borderlands criticism, in attempting to displace hegemonic, monologic cultural practices of the nation-state, ironically keeps the border "migrant" moving, en route. This post-national valorization of the "deterritorialized" or "liminal" status of the hybrid subject participates in a form of what I would describe as "movement discourse," a discourse that articulates the American faith in "mobility" as being ultimately redemptive and progressive. In his discussion of cross-border shopping in this volume, Bryce Traister points out how post-national borderlands criticism re-affirms American ideals of consumerist choice and reiterates some of the tenets of a corporate driven globalization. I argue that the borderlands form of movement discourse similarly participates in the reproduction of the American mythos of mobility, thus reaffirming ideologies of neo-individualism that act in the service of late-capitalist consumer decisionism.[1] Moreover, in a valorization of "routedness" over "rootedness," to use anthropologist James Clifford's useful distinction, the practices or narratives of border migrants that attempt to construct communal or subjective stability are either deconstructed or deemed the product of a colonized false-consciousness.[2] Ironically, then, contemporary forms of borderlands criticism tend to parallel the neo-imperialist exploitation of migrants; in both practices the immigrants' arrival is always deferred and the ways in which they might insist on arrival are pushed aside.

Most practitioners of borderlands criticism argue for an understanding of the border as an assault on the perception of a national culture as a "frozen worldview" that is both static and hegemonic. As José David Saldívar, the perhaps most influential practitioner of border theory, has put it, borderlands practice seeks "the deconstruction of the discourse of boundaries" where the "material hybridization of culture on the border" replaces the static version of culture (Saldívar 24–5). In *Border Matters,* Saldívar's influential formulation of borderlands criticism, he writes: "While I grew up believing in the truth of precepts that were available to me—the reality of space, borders, and nations—I now locate myself within a zone of dangerous crossings with new "centralities" that challenge dominant national centers of identity and culture" (19). Saldívar's definition of the borderlands emphasizes the disruption of what he describes as the "mainline U.S. *Bildung* through the creation of 'new migrant cultures'" (19). Thus, he follows Gloria Anzaldúa's lead who suggests in her landmark work *Borderlands/La Frontera* that the borderlands have become the new poststructural geographical trope that is populated by liminal figures, hybrid border-crossers, and migratory subjects as personified in the new borderlands subject, *la nueva mestiza.* This wish for intervention is based upon a valorization of "border crossings" that take the form of mobility, liminality, and hybridity.

D. Emily Hicks's articulation of her brand of borderlands theory, *Border Writing: The Multidimensional Text,* demonstrates a form of borderlands critique that is even more deeply indebted to poststructural theory. In her introduction, "Border Writing as Deterritorialization," Hicks announces that she intends to deploy the "migrant function" and to appropriate the "alterity" that comes from displacement and suffering. For Hicks, so-called "border writing" represents a "dispersal of the hegemonic order of a dominant culture" that "allows us to look in two directions simultaneously" (xxix). In a turn that the immigrant/migrant subject would probably find extraordinarily problematic, Hicks valorizes the *coyote* (profiteer and exploiter) as the "guide" to this space of "cultural, linguistic, and political deterritorialization" (xxxiv).[3]

Hicks's form of borderlands criticism describes migrant displacement as an "alternative to preserving the status quo" (3), in which "characters are freed from a single ordering or sequencing of reality through multidimensional or nonsynchronic memory" (4). Exploitation and repression are here transformed into operations of memory and perception to which the "solution" of an immaterial "freedom" can then be posited. Under this scheme, the regressive act of "reterritorialization" is defined as "the nostalgia for a previous ordering of desire . . . that attempts to fix people and objects in a structure" (112). Hicks associates this phenomenon with reactionary, imperialist desire. Under this paradigm, the displaced migrant worker who searches for a stable space in which to achieve her objective—material stability—is seen as simply enacting a colonized desire. Echoing Gloria Anzaldúa's definition of the borderlands as operating "in a constant state of transition," Hicks then condemns the migrant to circulate forever in his so-called "freedom," for "the destination of the border writer [and migrant] is to cross the border as many times as possible" (123). "Unblocked desire"—deterritorialized and "unfixed"—is ultimately valorized because it circumvents all structure and "order." In transforming the material effects of labor exploitation and displacement into the basis for a discursive practice, Hicks thus not only appropriates the migrant's "liminal" identity, but also dismisses his desire for arrival and stability in the national sphere by instituting permanent displacement.

Hicks's brand of borderlands theory has serious and far-reaching implications for the (de)construction of a realpolitik at the turn of the twenty-first century. This version of cultural studies locates struggle primarily within the cultural sphere, in which "culture" is defined mainly through a consumerist structure. In this view, social change and justice can be effected through the "transculturation" of the culturally dominant by entering into the production and consumption cycle, what Neil Larsen has called "consumptive production" (xiii). How such a "subversion" of the "culturally dominant" avoids simply participating in the cycle of consumerism, however, is not addressed.

The second manifestation of this type of borderlands politics is the creation of a form of romanticism in the guise of "radical subjectivity." Fredric Jameson has suggested that radical politics either foreground a communally oriented "triumph of the collectivity" or a more individualistic "liberation of the soul or spiritual body" (112). Locating all struggle and progress within the process of subject formation is ultimately little different from the self-absorbed individualism that undergirds most nineteenth- and twentieth-century movements of "dissent" in America. It posits an idealized self-determinism (performing in the cultural sphere) in ways that are alarmingly similar to an older American narrative that romanticizes displacement. In so doing, borderlands criticism risks replicating the master narratives of American nationalism that it purportedly seeks to deconstruct. A valorization of liminality (and the displacement that causes it), however, verges on inscribing the borderlands as the new frontier in a misguided celebration of mobility that replicates an older American mythos that conflates mobility, progress, and liberty.

This brings me to the third ramification of borderlands politics. By keeping the deterritorialized borderlands subject in a constant state of flux, the migrant remains isolated. In instituting permanent liminality as a political strategy, borderlands theory opens a space for the emergence of a form of culturalism (isolation and "resistant" consumerism) that may ultimately be as exclusive as Chicano *Movimiento*-era cultural nationalism of the 1950s–1970s. This danger has been most evident in the "Hispanicization" of a large number of Mexican Americans during the 1980s. It is well to remember that in its separatist mode, the Chicano movement abrogated its rightful claim to full participation in a larger public sphere and in the shaping a national culture. The *Moviemento*'s own exclusionary brand of identity politics alienated many who, left in a political vacuum, drifted to the right or chose to "practice" their politics through consumerism. Simultaneously, separatist politics were also instituted in the cultural sphere, where "resistance" largely meant not participating in the "mainstream." Both of these apolitical "culturalisms" are also distinctly operative in Hicks's form of post-structural/postcolonial borderlands theory.

Lastly, borderlands theory calls for what Hicks terms a "politics of non-identity." Race, ethnicity, and all other significant elements of identity formation are defined purely at the interstices of their interaction. If identity is in constant flux, we are giving into a form of self-erasure in which people's identities are seen as a set of discursive practices and interventions. The fear that drives this form of poststructural/postcolonial borderlands theory is the fear of fixity (the fear of being colonized) so that all attempts or desires for "arrival," all "institutions," all civic, national structures are seen as constraining. In rightfully challenging non-democratic, non-participatory versions of the nation-state, we have endorsed the creation of an immaterial, so-called "hybrid" culture rather than an inclusive national culture that calls for social justice.

It might be useful to turn our attention now to the ways in which the ideas of a "deterritorialized" and hybrid borderlands can also be used for distinctly neo-imperialist purposes. As even a cursory reading of migrant literature by writers such as Tomás Rivera, Luis Valdez, Ramón Perez, Ernesto Galarza, or Helena María Viramontes demonstrates, the migrant desires above all to arrive, to get home, to make roots. My critique cannot grant the migrant his desires, but it can acknowledge those desires as well as draw out the powerful critique of systemic forces that kept the prospects for arrival at the horizon's distant and unreachable end. Read in this way, migrant discourse, in part, helps to construct a complex critique of the ideologies inherent in borderlands narratives of dissent and to offer a renewed sense of *Americanismo.* The notion of the *Americano,* as envisioned by the Mexican-American generation of the 1930s and 1940s, insisted on a participatory subjectivity within the American nation-state. It affirmed the location of the public sphere within civic and institutional operation, and recognized the importance of "place," "citizenship," and "nationhood" in enacting radical democratic reform.

While I will not claim that at this time all Mexican migrant workers wished to become citizens nor that the Mexican American citizen enjoyed the full array of the prerogatives of citizenship, I argue that many longed for a form of stability and respect and self-determination that first-class, fully functioning citizenship bestows in principle (if not always in practice). Migrant texts bear out an understanding of the importance of "citizenship" that nationalist/separatist activists often dismissed. My use of the word *Americanismo* is thus used in contradistinction to the rather narrow conceptualization and practice of citizenship that the pervasive notion of "Americanism" denoted before being challenged and partially dismantled by the 1960s counterculture (including the Chicano Movement). Although Ernesto Galarza did not coin the term *Americanismo,* his work articulates its tenets: the desire to participate not only within the local community but also to impact and transform the national public sphere through a committed program of direct political participation on both the local and national levels. As an activist, Galarza (along with figures such as César Chávez and Tomás Rivera) practiced and advocated *Americanismo* in response to a very narrow and exclusionary view of the public sphere, democratic participation, and definition of citizenship held not only by mainstream Americans but also by advocates of separatist and nationalist agendas, and most recently, by advocates of radical culturalism that in my reading seek to institute a permanent instability as a form of social redress.

In his sociological work on migrant farmworkers who came to the United States under the Bracero Program (a program recruiting Mexicans as "guest workers" between 1942 and 1964), Galarza exposes the strategy by which "fluidity" is exploited by capital and displacement is reified as the quintessence of liberty and individualism. He shows that the border displaces and enables migrant disruption and exploitation. In *Merchants of Labor,* Galarza explains that the migrant's material experience exposes the American mythos of mobility as exploitative and ultimately immobilizing.[4] Migrant fluidity and mobility are seen not as liberating (as in a certain practice of border theory), but as the most exploitative form of displacement. Galarza posits that this mobility is deployed by agribusiness deliberately to de-stabilize the migrant and the migrant community in order to assure a "fluid labor pool."

With the introduction of the McCarran-Hartley Acts of 1950 and 1952, the deportation of *braceros* and undocumented workers became even easier. Under the 1950 rule, deportation was allowed on the merest suspicion of communist affiliation or "subversive" activity. This provision was expanded to cover almost any kind of action deemed threatening to the grower or agribusiness. The McCarran Act of 1952 went so far as to authorize the building of six detention camps for those suspected of opposition to agribusiness. The fear of deportation or detention made it very difficult for farm workers to unionize or even to demonstrate for better working conditions and wages.

Galarza's *Merchants of Labor* is an indictment of American capitalism, of its human rights violations, of the cynical use of the power to "fix" and "unfix," to detain and set free, and ultimately to enslave. Galarza puts this question to his reader: "Is this indentured alien—an almost perfect model of the economic man, an 'input factor' stripped of the political and social attributes that liberal democracy likes to ascribe to all human beings ideally—is this bracero the prototype of the production man of the future?" (16). Galarza's foundational premise is that movement for the migrant is not a choice. Rather, the migrant merely trades one national form of peonage for another. In recounting the serf-based economy of Mexico, Galarza sets up a comparison with the laissez faire system in the United States, leaving the reader to draw the conclusion that the serf/migrant is exploited by both systems. The migrant's search for *el pueblo libre* is thus doomed to failure, for the fictional space does not exist, cannot exist, when the systemic forces of capitalism and the revolving door are predicated against its existence.

In his autobiographical novel *Barrio Boy,* Galarza additionally insists that migration attempts to re-establish in the United States conditions for the migrant that allow for a stable communal and familial existence. Migration, writes Galarza, is the failure of roots. Throughout *Barrio Boy,* there is a palpable sense of loss and nostalgia for a life that revolves around the security of a homestead. In essence, Galarza's notion of *el pueblo libre,* which he formulates in *Barrio Boy,* can be read as a yearning by the mobile-weary migrant for a distinctively different model of the borderlands, one which promises stability rather than displacement.

As Galarza's protagonist Ernesto moves north, he finds that the organic space he has left existed only as an insulated space. Once displaced by the forces of the revolution, he discovers that to escape its disruptive influence means entering the pull of another set of disruptive forces north of the border. The *patrones* (land owners/agricultural employers), the *autoridades* (authorities), and the military that inevitably disrupt the stability of *el pueblo libre* cannot be escaped. They simply reappear, and the same systemic forces are replicated; they are just as circular, centralizing, and centrifugal. The migrant soon learns that stability is counterproductive to an industrialized market economy. Mobility is harnessed and scheduled. This is represented via the burgeoning railroad system in which Ernesto's uncles find work. As Ernesto and his family move north, transportation systems become more advanced and complex: from burro to stagecoach, to train and finally to the automobile. Paradoxically, however, more advanced transportation signals stricter industrial and economic organization, and thus a more restrictive environment.

Therefore, Galarza suggests that progress for his people lies not in fluidity or in increased mobility. Rather progress has its source in "arrival." For Galarza, a homestead does not equal a repressive domesticity. It provides the basis for community, for a stability in which self-determination can actually become reality. The site of *el pueblo libre* is not found in a utopian space, but is instead discovered through the very real mobilization of *comunidad* (community). Organization unites community, workers, domestics, and internationals alike. The idealized *pueblo* of his past, naturalized and monolithic in its make-up, broken-up and dispersed by the forces of "revolution" must be reconstructed, stabilized, and organized for the purposes of a unified front.

Like so many of Ernesto Galarza's projects, *Merchants of Labor* had a direct political impact. Passed out to every member of Congress as they debated the extension of the Bracero program, the book influenced enough members to defeat the resolution. Galarza's work historicizes the border and provides a strategy for constructing an arrival that insists on presence, stability, and political inclusion within the American nation-state. This realization has too often been placed under the rubric of "assimilation" and "selling out" when it should be recognized, as Galarza enacts his intervention, as direct political action that seeks to transform the American national culture. His work constitutes an effective alternative to both assimilationism and to separatist or culturalist politics.

While Houston Baker has usefully pointed out that "[f]ixity is a function of power" (202), a reading of Galarza's work adds to this insight that "movement" is also a function of power. The debates that have risen in California since 1986 with the passing of IRCA,[5] the overwhelming passage of Proposition 187 in 1994,[6] President Clinton's Operation Gatekeeper, the immigration bill that doubles the number of Border Patrol officers, and California's efforts to sue the federal government for the costs of "illegal immigration," only list the past decade's more reactionary attempts to use divisive national issues to reaffirm the power to control movement to make sure that the disenfranchised, the poor, the undocumented, and the person of color is kept "on the move."[7]

Events at Green Valley, a shanty town in southern California's La Costa, provide a more detailed example of the dominant culture's need to disperse the potentially communitarian community of color through forced movement. After the publication of a *Los Angeles Times* article on July 17, 1988, the residents of La Costa began calling in complaints about the shanty town even though, prior to the appearance of the article, many living in the area had not even been aware of its existence. The camp was on private property, which was owned by a rancher who used the workers in his fields. Nevertheless, the locals were able to have it closed by using the Health Department since the INS was unable to interfere as many of the residents of Green Valley were legally in the country.[8] The migrants and their attempts at community building remained trapped between neo-conservatives who sought to use them as a symbol of an "infiltration" of undesirables who must be "checked" and expelled, and liberal forces which, while attempting to protect them, also used them as a rallying point to counter the backlash of white discontent. As a resident of Green Valley said when he found out that the camp was to be destroyed:

> Why do the Americans detest the Latin Americans, the Mexicans? The North Americans invaded all of California. They colonized it . . . [Now] we come here to look for a job in what you could almost call our native land . . . We come in search of dollars. Your precious commodity. With one of your dollars we've earned our living for the entire day . . . You treat us like dogs. And that's what hurts because we're human, and hard workers because we're determined. We come here honorably. (qtd. in Chávez 111)

Buried within this rhetoric is an analysis of the economic exchange of which the migrant finds himself a victim. Aware that s/he is living in a colonized space, the migrant makes the case that the "precious commodity" of dollars, still a form of pure exchange, is not something which his search will net. Rather, s/he is put in the position of being apprised at much less than the "precious commodity." His place is to be chased away, moved at will, equivalent to the dollar only in the sense that he can be spent, saved, hoarded, or used.

The crime committed by the Green Valley residents was that, as a "transient" labor force, they attempted to create a community, a ramshackle realization of a permanent communal space. The goal of the demolition of Green Valley was to destroy the communal efforts of the migrant and to reassert jurisdiction; Green Valley's destruction reasserted the power to reify the migrant as a currency held by the dominant culture. What Green Valley "residents" asserted was their "arrival," a transgression that amounted to a claim of cultural and civic "citizenship." But this was an act that their "American" neighbors could not allow.

Thus issues of "movement" should also be understood also as involving questions of "access." The notion of access denotes the possibility of entry into a particular institution; consequently the desire for access, on the migrant's/subaltern's part, has an "objective." But this is exactly what is denied to the migrants by the "power-brokers," the interests of agribusiness, and corporate America—and ironically, also by poststructural/postcolonial borderlands theorists. As Galarza's work shows, the sort of unfixity that border theorists like Hicks find so liberating can be just as fixating and disempowering as any set of laws or practices that limits minority/migrant abilities to participate within the democratic process. The efforts to deny the migrant "place" and stability suggest the very real threat that the formation of community poses to the powerbrokers of American politics and the economy.

The need to keep forcing American studies and cultural studies to take into account the Mexican American narrative is imperative, no doubt. But should the phenomenon of migrant liminality be universalized without the recognition that this status is a result of exploitation and material suffering? Many border migrant narratives advocate a renewed sense of

Americanismo, which recognizes the importance of "place," "citizenship," and the role "nationhood" plays in enacting radical democratic reform. They also recognize that displacement is the source of material struggle and needs to be addressed as such, rather than "harnessed" as a utilizable effect in the service of de-stabilizing identity. This critical lesson of migrant and border experience calls not for a release from an "outdated" notion of the American nation-state, but instead insists, in *Americano* fashion, upon inclusion within a revised construction of the nation-state conceived as inclusive and participatory. In fact, border/migrant narratives exhibit an *Americano* yearning for arrival and participation within the nation-state by attempting to radically restructure a national culture through an insistence on equal and full participation that depends upon a communalist model of direct action politics.

In destabilizing and deconstructing the idea of the national sphere, or of the "nation" itself, have we in essence turned a blind eye to the reality of nation, the necessity of a "genuine" and "legitimate" place from and in which we might practice our rights to full inclusion and participation? One thing that the New Left learned from its predecessors, the Old Left, according to Stanley Aronowitz, was to ground their progressive leftism in native ideological currents. Aronowitz points specifically to documents such as *The Port Huron Statement* that articulated its egalitarian agenda through a democratic discourse located and practiced within the American nation-state. Chantal Mouffe and Ernesto Laclau have similarly advocated egalitarian redress through the practice of radical democracy that emphasizes communal stability and the creation of an inclusively derived national culture through participatory reforms.[9] Thus, poststructural reading practices of the border are useful in what they uncover about the exploitative and authoritarian nature of market capitalism, the reality of forced mobility or the enforcement of an ever fluid labor pool. But this recognition should actually give credence to the idea of "nation" and "citizen" as necessary vital constructs through which to enact egalitarian reform. The question then is, where do we want to enact progress and where do we need to locate such activity?

I suggest that what is most important is the immigrants' effort to claim a stable communal space from which to practice subsequently full participation. If these efforts are being actively persecuted by nativist and reactionary forces, it is urgent that rather than deconstruct the language and the very possibility of stable identity and place, those of us desiring to participate within an *Americano* project rehabilitate the notion of community, national identity, and civic participation. Doing so gestures toward a recognition of social justice and equal access.

This project, in part, requires that we recognize that the valorization of displacement in poststructural borderlands theory render the migrant's experience an abstracted phenomenon, a migratory reading strategy, that ultimately enacts a tragic discursive exploitation. Practicing these forms of theoretical displacements creates an essentialized "migrant-function" that the critic then appropriates for its double-consciousness–giving alterity. The migrant becomes his/her "disguise," allowing him/her entry into the liminal space of the "marginal." Rafael Pérez-Torres has eloquently described this process as follows:

> From those who can least afford to be exploited, a migratory reading steals strategies from the lived practices of the dispossessed. This intellectual exploitation tries to profit from the improvisation and negotiation born of necessity in the hard and hostile world of transnational survival . . . [Yet] the profit born out of this exploitation is meant for those from whom it steals. (170)

And yet Pérez-Torres's insight is unsatisfactory because the migrant of the past cannot receive any such "profit." Nor can a decontextualized, abstracted application of migratory reading practices, the "migrant function," be productive in exploring the dimensions and avenues of turn-of-the-twenty-first-century *Americano* political action. Rather, what should be understood as central to the migrant experience is the desire for arrival, the migrant's social awareness, and the political action that emerged directly from material experience. This social awareness and its political dimensions suggest a model for constructing a progressive *Americano* politics that will result in political, social, and cultural changes within the American sociopolitical landscape.[10]

I agree with Pérez-Torres's contention that, "one must insist upon the fact of deterritorialization as a historically grounded, painful, and often coerced dislocation" (151). The "gain" of such forced movement may in the end be the acquisition of a tool for deconstructing master narratives that serves perhaps as a poststructural strategy for "the dissolution of ordering systems." But before such effects can be exploited, we must recognize that the migrant's desire to arrive also has tremendous import for creating a politics that is responsive to the formation of community and political participation. "Arrival" is central for powering an agenda based on both a constitutional and cultural understanding of "citizenship" that stresses access and inclusion. The concept and practice of "citizenship" in a national entity should thus be expanded, not dismissed.

What both the Green Valley case and Galarza's work show is that the border and its power to deny access to the political and legal realm by keeping the migrant/minority border subject in a constant state of unfixity must be directly confronted. What is at stake is not merely the tensions or antagonisms inherent in a particular discursive site nor should these activities be seen as charged points for "resistance." Thus, I must differ with José

David Saldívar's contention that poststructural deployment of hybrid techno-cultural strategies and productions have "utopian uses as new forms of resistance and struggle" (35). Ultimately, these cultural strategies do not assert "arrival" in the same way that Galarza's protagonists and the Green Valley residents desired. By practicing a form of border theory that focuses on "in-betweenness" (Saldívar 56) rather than the *Americano* emphasis of "arrival," we sell short the potential of democratic constructivism, of radical democracy, supplanting it with a radical "culturalism" that is too easily co-opted by the market and that tragically replaces the "political" sphere from which we might launch an insistence on citizenship and its material practice and rights.

I'm reminded of living in California in 1994 during the highest pitch of nativism in decades and being confronted with the absurdity of a Taco Bell commercial urging viewers to make a run for the border. It was aired right after a commercial paid for by Pete Wilson and the Republican Party depicting undocumented workers running across the border en masse. That juxtaposition speaks volumes about the economic and political realities of the border and the way that American capital so easily co-opts the notion of hybridity as a slickly packaged version of innocuous multiculturalism. So, too, can the discourse of "culturalism" be co-opted by reactionary forces that circumvent what George Will refers to as "material distribution," by instead launching "a comprehensive attack on the culture of poverty, with measures ranging from welfare reform . . . to school choice" (qtd. in Will A11). A "borderlands" understood through the experience of the material border in all its repressive, exploitative power resists its appropriation as either a nationalist symbol or a commodified object used to sell some new version of the melting pot. It is the awareness of the "real" border, which keeps its social and cultural critique from becoming a set of deconstructive practices that are ultimately devoid of a realpolitik.

To articulate a question that has inspired this essay: Is the displacement of Mexican/Mexican American subjects by the interpolation of the border between Mexico and the United States in 1848 somehow replicated in borderlands theoretical practices? If so, then displaced peoples are again displaced, this time textually. No longer in the margins, they have now been relegated to the footnotes of new narratives, their yearning for arrival, to paraphrase Tomás Rivera, deferred by a too facile valorization of liminality. Ultimately, distinguishing between the border and the borderlands means facing that the U.S.-Mexico border represents the all too real repressive power of the state to define, to limit, to imprison, and even to kill. While it is a necessary move to point out resistance to this power to define subjectivity, we might remember that the border as a site for repressive state power remains intact.

I want to conclude by demonstrating the importance of making the border distinctive from the borderlands, of acknowledging the reality while utilizing the liminality. I will quote from a well-known passage in Tomás Rivera's poignant novel of migrant life, *y no se lo trago la tierra* (*and the earth did not devour him*). In it, we hear the voice of a woman who has been standing in the back of a truck trailer as she and a group of other migrants travel north to Minnesota:

> When we arrive, when we arrive, the real truth is that I'm tired of arriving. Arriving and leaving, it's the same thing because we no sooner arrive and . . . the real truth of the matter . . . I'm tired of arriving. I really should say when we don't arrive because that's the real truth. We never arrive. (145)

We can interpret this passage in several ways, and each interpretation can be used for a specific purpose. We can read this as a commentary on the impossibility of reaching meaning itself, a commentary on deconstructive reading practices in which "truth" is forever deferred. Relatedly, the passage can be seen as a representation of the migrant's liminal position through which such deference allows for recognition of the discursive nature of social, cultural, and ideological structures, their a priori status exposed as pure construction. Read another way, the passage becomes the eternal cry of the proletariat, or more locally, the state of American labor, displaced and mobile, chasing an elusive American dream constructed to keep the postindustrial subject on the move. It might even be read as a commentary on the nature of postwar progressivism written by a somewhat disillusioned Chicano liberal at the end of *el Movimiento.* These interpretations are within their own context valid and are in fact quite useful in redressing social injustice or in positing both an unfixed subject position and a destabilizing reading strategy. But the "real truth," to quote the migrant worker, is that she is tired, physically and spiritually. Her plaintive, internal cry is based on her very real suffering and that needs to be recognized or we risk, as critics and readers, re-enacting the dynamics of her displacement, rendering her pain and suffering irrelevant, relegating her once again to the margins, while paradoxically elevating and utilizing her liminal subjective position as a critical function. That, I argue, is an inadequate response.

Notes

1. See Frederic Jameson.
2. See James Clifford's *Routes,* which borrows the routes/roots distinction from Paul Gilroy's *The Black Atlantic.*

3. Hicks's valorization of the generic "coyote" differs markedly from Ursula Biemann's more differentiated theorization of a certain kind of people smuggler. In her article in this volume, Biemann describes Concha, a border crossing guide who helps pregnant Mexican, Central, and South American women across the border, but charges little for her services.
4. See Galarza's *Merchants of Labor* and *Tragedy at Chualar* for prime examples of his work in documenting the material suffering of migrant workers and braceros and its systemic causes. See also James Cockcroft; David Montejano.
5. On the Immigration Reform and Control Act of 1986, see Cockcroft and Bean's *Undocumented Migration to the United States.*
6. For the implications of Proposition 187, see Tomás Rivera Center, *California School District Administrators Speak to Proposition 187.* For a discussion of the overall implications of anti-immigration legislation see Mills.
7. See Charles Gallagher's "White Reconstruction in the University," in which he describes a renewed emphasis on re-constructing "whiteness" as an ethnicity based on a "persecutionist" model that dehistoricizes as it attempts to protect Anglo economic interests.
8. For a full discussion on the ramifications of Green Valley see Leo Chávez's *Shadowed Lives.*
9. See Laclau and Mouffe, *Hegemony and Socialist Strategy: Towards a Radical Democratic Politics* and Aronowitz, *The Death and Rebirth of American Radicalism.*
10. Yet another example of this sort of theoretical reduction of the migrant to a function can be found in Lisa Lowe's use of the migrant-function in her analysis of Peter Wang's film *A Great Wall.* Pérez-Torres quotes Lowe in the following way:

    [Wang] performs a filmic "migration" by shuttling between the various cultural spaces; we are left, by the end of the movie, with a sense of culture as dynamic and open, the result of a continual process of visiting and revisiting a plurality of cultural sites. (Pérez-Torres 131)

Works Cited

Anzaldúa, Gloria. *Borderlands/La Frontera.* San Francisco, CA: Aunt Lute Books, 1987.

Aronowitz, Stanley. *The Death and Rebirth of American Radicalism.* London: Routledge, 1996.

Baker, Houston. *Blues, Ideology and Afro-American Literature.* Chicago, IL: University of Chicago Press, 1984.

Bean, Frank D. and Barry Edmonston. *Undocumented Migration to the United States: Irca and the Experience of the 1980s.* Washington, D.C.: Urban Institute Press, 1990.

Chávez, Leo. *Shadowed Lives: Undocumented Immigrants in American Society.* New York: Harcourt, Brace, Jovanovich, 1992.

Clifford, James. *Routes: Travel and Translation in the Late Twentieth Century.* Cambridge, MA: Harvard University Press, 1997.

Cockcroft, James. *Outlaws in the Promised Land: Mexican Immigration Workers and America's Future.* New York: Grove Press, 1986.

Galarza, Ernesto. *Barrio Boy.* Notre Dame, IN: Notre Dame University Press, 1971.

———. *Merchants of Labor.* Santa Barbara, CA: McNally and Lofton, 1964.

———. *Tragedy at Chualar.* Santa Barbara, CA: McNally and Loftin, 1977.

Gallagher, Charles A. "White Reconstruction in the University," *Socialist Review* (1996): 165–183.

Gilroy, Paul. *The Black Atlantic: Modernity and Double Consciousness.* Cambridge, MA: Harvard University Press, 1993.

A Great Wall. (Videorecording and Screenplay) Co-Written by Shirley Sun, Produced by Shirley Sun, Written & directed by Peter Wang, Beverly Hills, CA: Pacific Arts Video, 1987.

Gutiérrez, David G., ed. *Between Two Worlds: Mexican Immigrants in the United States.* 15. Wilmington, DE: Scholarly Resources, 1996.

Hayden, Tom. *The Port Huron Statement: The Founding Manifesto of Students for a Democratic Society.* New York: Charles H. Kerr Publishing Co, 1990.

Hicks, Emily D. *Border Writing: The Multidimensional Text.* Minneapolis, MN: University of Minnesota Press, 1991.

Holley, Joe. "Border Problem: Sheriff Authorized to Arrest Immigrants," *San Antonio Express-News* (February 5, 2000): 1A.

Jameson, Frederic. *Postmodernism, or, the Logic of Late Capitalism.* Durham, NC: Duke University Press, 1991

Kerouac, Jack. *On the Road.* New York: Viking Press, 1957.

Laclau, Ernesto and Chantal Mouffe. *Hegemony and Socialist Strategy: Towards a Radical Democratic Politics.* New York: Verso, 1985.

Larsen, Neil. "Foreword" *Border Writing: The Multidimensional Text.* Emily Hicks. Minneapolis, MD: University of Minnesota Press, 1991.

Limón, José. *American Encounters: Greater Mexico, the United States, and the Erotics of Culture.* Boston, MA: Beacon Press, 1998.

Maciel, David R. and María Herrera-Sobek, eds. *Culture Across Borders: Mexican Immigration and Popular Culture.* Tucson, AZ: University of Arizona Press, 1998.

Mills, Nicolaus, ed. *Arguing Immigration: The Debate Over the Changing Face of America.* New York: Touchstone, 1994.

Montejano, David. *Anglos and Mexicans in the Making of Texas.* Austin, TX: University of Texas Press, 1987.

Mouffe, Chantal, ed. *Dimensions of Radical Democracy: Pluralism, Citizenship, Community.* London: Verso, 1992.

Paredes, Américo. *With a Pistol in His Hand.* Austin, TX: Arte Público Press, 1958.

Pérez-Torres, Rafael. *Movements in Chicano Poetry: Against Myths, Against Margins.* Cambridge: Cambridge University Press, 1995.

Reza, H. G. "Home Cooking Among the Hooches of La Costa: Two Restaurants Provide Familiar Flavors for the Residents of Hidden Migrant Camp," *Los Angeles Times* (July, 17 1988): B1.

Rivera, Tomás. *Y no se lo trago la tierra/and the earth did not devour him.* Houston, TX: Arte Público Press, 1987.
Saldívar, José David. *Border Matters: Remapping American Cultural Studies.* Berkeley, CA: University of California Press, 1997.
Tomás Rivera Center. *California School District Administrators Speak to Proposition 187: TRC Survey.* Claremont, CA: Tomás Rivera Center, 1994.
Will, George. "Bradley Playing Out the Clock," *Sunday Herald Times* (March 5, 2000): A11.

Reading across Diaspora

Chinese and Mexican Undocumented Immigration across U.S. Land Borders[1]

Claudia Sadowski-Smith

In the aftermath of the September 11, 2001 terrorist attacks on the World Trade Center and the Pentagon, the 5,000-mile U.S. border with Canada has been elevated to an unprecedented place of prominence in the American public consciousness. The closing of crossing points, long delays for cross-border traffic, and detailed searches at ports of entry in the wake of the attacks resembled similar types of systematic crackdown at the 2,000-mile U.S.-Mexico border throughout the 1990s. These types of border enforcement also put a temporary end to efforts for the creation of a common "open and seamless border" between Canada and the United States. First articulated during the summer of 2001, proposals to ease U.S.-Canada border restrictions under a "NAFTA-plus" plan were couched in appeals to the long and peaceful history of "the world's largest undefended border" that separates two largely similar neighbors (Walker A1).

All this suddenly changed when early investigative reports began to suggest that some of the terrorists might have illegally entered the United States via its northern border. Scores of pundits singled out for blame Canada's "permissive" refugee laws and its more liberal non-visa requirement policies with a variety of countries, among them nations that the United States opposes (Dorning A5).[2] Even though we soon found out that all of the nineteen terrorists arrived legally on tourist, business, and student visas, the mere repetition of arguments that Canada has become a "pathway for terrorists" have sufficed to make it so. The rhetoric connecting terrorism to "foreigners" who abuse differences in national immigration legislation could then also be used to pressure Canada into changing its policies and to articulate demands for more U.S. Border Patrol agents and stricter immigration policies in the United States itself.

The link between illegal activities at the U.S.-Canada border and terrorism directed at the United States has a much longer history. It dates back to at least 1997 when the Palestinian Gazi Ibrahim Abu Mezer was prevented from crossing the border with bomb-making material designed to blow up the New York subway system. On April 14, 1998 Congress launched its very first hearing on the northern border, aptly entitled "Law Enforcement Problems at the Border between the United States and Canada." While there was surprisingly little mention of terrorism, several witnesses emphasized the rise of cross-border drug smuggling and of undocumented Asian (especially Chinese) immigration since the late 1980s ("Law Enforcement" 6).

Then, in December 1999, the Algerian Ahmed Ressam was captured at Port Angeles with a false Canadian passport. Ressam had tried to cross into the United States to blow up the Los Angeles airport during the Millennium Celebrations. He later admitted that he had been trained in the art of bomb making in camps connected to Osama Bin Laden's network. But again, initial concerns over what appeared to be Middle Eastern terrorist activity spilling from Canada into the United States quickly gave way to more familiar narratives about "illegal" immigration. After the apprehension of other improperly documented Jordanian immigrants at a Washington State border crossing a few days later, the news media began to stress the largely unmarked and unprotected nature of the U.S.-Canada frontier. To illustrate just how unsafe the world's longest undefended international border had become, viewers were presented with previously unaired high-tech surveillance footage of an understaffed Border Patrol capturing "illegal" immigrants, and the media also started showing reports about rising illegal immigration from China.

The repeated connection between Middle Eastern terrorist threats and unauthorized U.S.-Canada border crossings complicates scholarly and popular discourses that have, throughout the 1980s and 1990s, identified border porosity and undocumented immigration exclusively with the U.S. Southwest.[3] Border studies' primary focus on Mexican immigration and the U.S. Southwest has as much to do with the field's origin in Chicana/o studies as with the very different histories and cultural meanings of the Mexican and Canadian borderlands in relation to U.S. attempts at domination, which Bryce Traister vividly evokes in his contribution to this volume.[4] Chicana/o studies originally claimed the U.S. Southwest as a geography from which to exact resistance to exclusive discourses of U.S. nationalism.[5] Its recent expansion across U.S. borders has shifted the focus from Chicana/os' experiences of biculturality and Mexican border crossings to transnational cultural practices within a "Hispanic" diaspora.[6] This work stresses Latin, Central, and South American immigrants' ability to maintain spatially extended relationships between the United States and their countries or areas of origin, which undermines both normative ideologies of U.S. nationalism and oppressive practices of the U.S. nation-state.

As I argue in this essay, the new emphasis on porosity and illegality at the U.S.-Canada border also opens up opportunities for cross-cultural work that moves beyond and below such currently available conceptualizations of U.S. diasporic immigrant formations as bounded by a common culture, language, geographical home, and/or descent. The essay focuses on similarities in the diasporic "travels" of two groups of undocumented immigrants across the U.S. borders with Mexico and Canada: the more than one million Mexican *indocumentados* who have come to the United States every year since the late 1980s and a comparatively much smaller constituency of up to 100,000 undocumented Chinese (P. Smith x). Unlike other immigrants who are defined as students, professionals, or entrepreneurs within juridical-political categories that cut across culturally and ethnically constituted identities, these groups of *indocumentados* have had to similarly negotiate accelerated border militarization, the passage of exclusionist immigration restrictions, and more massive border policing throughout the 1990s and into the twenty-first century.

Thus, border militarization does not simply serve the interests of the U.S. nation-state in shoring up its national boundaries to racially diverse immigration, as has been suggested in much diaspora and transnational scholarship. The increasing distinction between culturally and politically constituted immigrant identities—formalized within differentiated border crossing regimes—also needs to be understood in the context of joint corporate and national/local efforts to accommodate ongoing processes of globalization. A more complicated U.S. policy of barring some people while allowing a modest number of migrants, sometimes of the same national or ethnic background to enter and permanently stay in the United States can be seen as one particular response of the nation-state to the transnationalization of labor and production. The deepening of economic, political, and cultural ties with countries—like China and Mexico—that have received U.S. production facilities has also produced the necessary conditions for large-scale out-migration that takes *both* documented and undocumented forms. This essay begins by weaving a historical narrative of undocumented forms of immigration from China and Mexico, and then focuses on exploring their contemporary convergence at reinforced U.S. borders. I propose that such a cross-cultural inquiry not only helps to make various groups of border crossers more visible to one another, but that it may also contribute to articulations of new and as yet unformalized forms of cross-cultural agency in an increasingly globalized context.

Forbidden Workers[7]

One important step toward the articulation of contemporary cross-cultural interactions is to trace the somewhat understudied similarities in the histories of Chinese and Mexican undocumented immigrants. After all, the confluence of various legal, political, and economic factors at the turn of the twentieth century created not only the Mexican *indocumentado* but also the Chinese undocumented border crosser. In fact, the 1882 Chinese Exclusion Act first institutionalized the very category of the "illegal" immigrant and also identified it with citizens from a particular country of origin in ways that conflated certain nationalities with notions of ethnically defined "otherness." The opening of Chinese ports for commerce and the liberalization of anti-emigration laws in the nineteenth century encouraged out-migration to several places, among them the United States, where Chinese worked in gold mines or on railroad expansion projects. Only three decades after its beginning in the 1850s, however, various Chinese Exclusion Acts suspended the immigration of Chinese laborers (thus virtually barring the entrance of all Chinese) and banned the naturalization of Chinese already in the United States. Forced to enter or re-enter the United States "illegally," over the next three decades Chinese immigrants transformed the then largely unguarded land borders with Canada and Mexico into the second most popular ports of entry after New York (Lee *At America's Gates*).

New historical scholarship has unearthed the fact that the earliest and most numerous undocumented border crossings occurred at the Canadian border. Until 1920, this boundary was not supervised by custom and immigration officials, and its outline had not even been firmly established in multilateral agreements (McIlwraith 54). While no official statistics exist, historians assume that at least a few thousand Chinese entered the country via Canada between the 1890s and the early 1900s (Lee *At America's Gates*). Chinese chose this site because Canada was, until the passage of its own Exclusion Act in 1923, much more lenient in admitting Chinese.[8] The leveling of head taxes on Chinese immigration, introduced in 1885, only made the entry into Canada, but not U.S. immigration via the U.S.-Canada land border, a less attractive option. Canadian immigration law actually even allowed those Chinese destined for the United States to remain in the dominion for ninety days without paying the required head tax (Lee *At America's Gates*). The majority of undocumented Chinese crossed the border in the Puget Sound area and in sites near Detroit and Buffalo, from where they would often be transported by smuggling networks to Boston, Chicago, and New York City.

After the passage of a 1898 U.S. law, fraudulent claims to U.S. citizenship on the basis of the so-called paper son system began to surpass undocumented border crossings as the most popular means of violating the Chinese Exclusion Act. In addition, under continual pressure from the United States, in the early twentieth century Canadian institutions began to cooperate with U.S. authorities in controlling undocumented immigration across the U.S.-Canada border (Lee "Crossing Borders"). Consequently, Chinese immigrants largely moved their points of entry from the Canadian to the Mexican frontier, which was already being transformed into a gateway for undocumented Chinese and Japanese as well as for the comparatively much larger number of those Europeans who were unable to pass inspection at Ellis Island.[9]

In comparison to Canadian authorities who agreed to support U.S. attempts at enforcing its exclusionary immigration laws against Chinese, Mexico rarely cooperated with U.S. agencies because it was at that time more interested in recruiting immigrants to support its ongoing modernization project (Lee *At America's Gates*). Whereas periodic outbreaks of racial hostility against Chinese in Mexico accelerated Chinese out-migration to the United States (especially during the Mexican Revolution), the surge in anti-immigrant sentiment did not translate into the passage of immigration restrictions as it did in the United States and later Canada. By 1906, reluctant cooperation on the official level and outright hostility toward U.S. immigration officials by Mexican authorities helped to elevate the Mexican border into the most important gateway for Chinese immigrants at the time.

Since official diplomacy and practical cooperation were largely unavailable at the southern frontier, in this region Chinese exclusion instead became enforced by means of border policing. As a consequence, the nature of the hitherto largely perfunctory (and almost invisible) line of international division between the United States and Mexico began to change dramatically. Up to this time, the two countries had not been separated by a formal barricade. Crossers only needed to report to a United States Port of Entry for inspection by the Customs Service, whose so-called line riders also patrolled the frontier and controlled undocumented immigration. In 1903, however, the line riders' responsibilities were assumed by the Bureau of Immigration with the creation of a Chinese Division and the appointment of officers to positions as Chinese Inspectors.

Around the same time, the Immigration Service also began recruiting officers as mounted watchmen and inspectors. A small group of mounted inspectors, later known as border mounted guards, was formed in El Paso, Texas, and soon extended its field of operation to New Mexico and Arizona ("A Brief"). Travel restrictions and the assignment of troops along the borders during World War I greatly reduced the influx of undocumented Chinese, and it largely came to a halt in 1916. The termination of steamship passenger service between Asia and Mexico, in particular, cut off those who intended to enter the United States via Mexican ports so that immigration into the United States effectively ended. Instead, out-migration to China increased: Between 1909 and 1939 more than 10,000 Chinese left the United States than entered it (Chun 175).[10]

While focusing its attention on curbing the immigration of an estimated 60,000 undocumented Chinese in the first three decades of the twentieth century, the Immigration Service pursued a policy of benign neglect toward the estimated 1.4 million of Mexicans entering the United States via the Southwestern border (Lee *At America's Gates*). During the early 1880s, Mexican movement across the U.S.-Mexico border was constant and uninhibited. Mexicans became the immigrant laborers of choice, largely also because of restrictions against Chinese (and, in 1907 after the passage of the Gentlemen's Agreement, against Japanese). Once created with respect to Chinese, however, the trope of the *indocumentado* could just as easily be applied to other groups, and border enforcement shifted to Mexican undocumented immigration. One indication of what was to come was the renaming of Chinese Inspectors (who had been appointed to curb Chinese immigration) into "Immigration Inspectors."

Largely tolerated before, Mexican immigration first became regulated in the 1920s. Despite an 1885 law restricting the importation of contract workers, greater numbers of Mexicans were informally recruited by U.S. employers to work in southwestern agriculture. Before enactment of the Immigration Act of 1917, few of the restrictions for overseas immigrants were applicable to Canadian and Mexican citizens, who were free to enter in unlimited numbers. The passage of the 1917 Act, however, expanded the imposition of literacy tests and of an eight-dollar head tax to Canadian and Mexican citizens, and thus declared most of the Mexican temporary labor ineligible for legal admission. No longer able to enter the United States on six-month work permits, the majority of Mexican seasonal laborers thus began coming "illegally." After the passage of the 1921 and 1924 Immigration Quota Acts, which established the nation's first limits on the number of people legally admitted every year, Mexican immigrants soon became the most regulated group of border crossers at the southern frontier. While the Border Patrol was formally created in 1924 to respond to Asian and European immigration, its institutionalization coincided with the sudden influx of more numerous Mexican immigrants who left the country because of deteriorating conditions in the aftermath of the Mexican revolution.

The Border Patrol took over the duties of the Immigration Service officers in the Chinese division as well as those of the former Customs Patrol at the U.S.-Mexico border and also drew on U.S. and Canadian law enforcement experiences of controlling the U.S.-Canada frontier. The memoirs of Clifford Perkins, one of the first Immigration Inspectors, indicate that the Border Patrol borrowed heavily from Canadian institutions charged with the enforcement of exclusionist U.S. (and later Canadian) immigration laws. Perkins remembers that the Border Patrol's uniform was initially modeled after what was then the Canadian Mounted Police (today's Canadian Royal Mounted Police) and that their members were contacted to share their experiences in enforcing exclusionary immigration legislation at the northern U.S. boundary (Perkins 90–91). Only three short years after the creation of the Border Patrol, its operations were extended from the U.S.-Mexico border to also cover Florida and the Canadian border (Perkins 91).

Theories of Border Crossings and Diaspora

Despite the role of Chinese immigrants as predecessors of the Mexican *indocumentado,* the movement of Mexicans into the United States has served as the preferred model for theories about undocumented U.S. border crossings. This framework is grounded in a view of Mexican undocumented immigration as having been relatively continuous since its origins in the 1920s, yet marked by various "peaks" in the mid-1960s (after the end of the Bracero Accord) and by a significant surge since the 1980s. The Bracero Program, which was in place between 1942 and 1964, initiated large-scale labor migration from Mexico to the United States. Between the termination of the Bracero Accord and the mid-1980s, the U.S. policy of active labor recruitment shifted to one of passive labor acceptance in the form of more massive undocumented entry.[11]

Various U.S. policies of the mid-1980s began to encourage formerly undocumented migrants to remain in the United States rather than shuttle back and forth across the border, and, as a result, the number of "illegal" migrant apprehensions temporarily dropped. Since 1989, however, undocumented immigration has again risen sharply and grown to over a million people by 1996 (U.S. Immigration n. pag.). The majority comes from the central Mexican states of Guanajuato, Michoacan, Jalisco, and Oaxaca (R. Gutiérrez 211). Mexican Americans and Mexicans who live in the United States are now commonly considered members of a modern diaspora. As Carlos Gonzáles Gutiérrez has put it, "they constitute a minority ethnic group of migrant origin which maintains sentimental or material links with its land of origin" (545).

How might our thinking about U.S. border crossings as inextricably linked to a Mexican (and increasingly a "Hispanic") diaspora change if we looked comparatively at Chinese undocumented immigration? Having originally been integrated into notions of a pan-ethnic Asian American identity and historiography, the study of Chinese immigration has recently become re-oriented to a more global perspective. According to historian Gordon H. Chang, new diasporic and transnational approaches to Chinese immigration have further problematized notions of an "authentic" shared cultural background and of a common ethnic identity. Chang notes that the about fifty million people who can be considered products of Chinese migration to over 100 countries since the fifteenth century are immensely diverse and include, for example, Taiwanese and Vietnamese Chinese who speak various different dialects, such as Mandarin or Cantonese (135–136).

Similar to worldwide Chinese migration, Chinese immigration to the United States has not been a continuous process. Instead, it has been characterized by two temporally distinct waves of immigration at the turn of the twentieth and twenty-first century, respectively. If Chinese Exclusion Acts first created undocumented Chinese immigrants, their repeal in 1943 was not accompanied by a large surge in migration.[12] Chinese immigration (both legal and undocumented) has only again become significant since the mid- and late 1980s (Massey xiii). In 1998, Chinese constituted the second largest group of immigrants (behind Mexicans) who were legally permitted to enter the United States ("Where" 1A). In general, the majority of contemporary Asian immigrants are (often ethnically Chinese) members of professional and elite backgrounds from Taiwan, Singapore, and Hong Kong who enter the United States legally. At the same time, however, up to 100,000 undocumented Chinese, most of whom are peasants from the Fuzhou area in the northeast coast of Fujian Province, have begun to enter the United States illegally every year (P. Smith x). As Douglas Massey has put it, although migratory flows have been most numerous from countries in Mexico, Latin, Central, and South America, the latent potential for immigration is greatest in China, which has a large population and where the forces that initiate and sustain international migration have only just begun to operate (xiii). Today, Chinese immigrants enter the United States either via U.S. airports or via the borders with Canada and Mexico (Chin 49). Thus, Chinese immigrants no longer alternate with Mexican undocumented migrants but rather overlap and often also converge spatially at U.S. land borders.

After the capture of several Chinese ships by U.S. authorities in the early 1990s, Chinese cartels began to employ more indirect routes from China to the United States. As these routes involve passage through Mexico (and sometimes other Latin, Central, and South American countries), immigrants are also required to clandestinely cross a U.S. land border (Kwong 79). Under the cooperation of human cargo smugglers—Chinese "snakeheads" and Mexican "coyotes"—the U.S.-Mexico border has become one of the major transit points for Chinese immigration (Rotella 72–73). In an interesting reversal of history, many Chinese have more recently also moved their entry points from the Mexican to the Canadian border.

Although contemporary Chinese and Mexican undocumented immigration draws on surprisingly similar historical precedents, their convergence at

U.S. land borders and the dramatic growth in their numbers suggests not simply a continuation of past processes, but rather a clear qualitative difference. Conceptions of the distinctive nature of contemporary U.S. immigration from so-called Third World countries have, however, tended to theorize a different kind of change. Much of the anti-nationalist agency formerly attributed to members of U.S. racialized groups or immigrants, such as Chicana/o and Mexican borderlanders, has been relocated into diasporic formations, like Hispanic, Latin, and Mexican ethnoscapes. This shift in focus stresses immigrants' ability to maintain spatially extended relationships between the United States and their respective sending countries or areas of origin in ways that undermine both normative ideologies of U.S. nationalism and oppressive practices of the U.S. nation-state. A growing number of people are identified as leading dual lives, speaking two languages, keeping homes in two countries, and sustaining regular contact across borders.

This view of diasporic subjects as agents of de-nationalization has also enabled comparative work on various different ethnoscapes, most notably articulated in anthropologist James Clifford's concept of "travelling cultures" and in Arjun Appadurai's notion of postnational "diasporic public spheres." Similar cross-overs are also (if less directly) acknowledged in the way that influential U.S. ethnic studies work has modeled itself after Paul Gilroy's scholarship on the African diaspora. As they are more interested in the *products* of cultural flows than in the process of migration, these influential theories have associated today's large scale movements of people with other forms of de-nationalization, such as the spread of information, the emergence of global media, and improved access to travel and communication technologies. All of these forces are said to contribute to the ongoing assault of various cultural forms on the U.S. nation-state.

In advancing a comparative theory of "travelling cultures," James Clifford, a pioneer of diaspora studies, has proposed "the border" as a place/metaphor of crossing and as a term of translation for comparative work. Clifford regards the practices of border crossing/immigration and experiences of diaspora as two closely related aspects of emerging transnational identities. He begins his chapter on "Diasporas" by discussing Roger Rouse's notion of the "transnational migrant circuit." In conjunction with other ethnographies of Mexican immigration, Rouse's work has helped to shift the focus from Mexicans' roles as laborers in the United States to their experiences as constituents of transnational entities that maintain relationships between Mexico and the United States. Clifford argues that while other kinds of diasporic formations bridge longer distances and sometimes also connect multiple communities, Mexican practices of border crossing nevertheless exhibit important diasporic dimensions. They help Mexican immigrants maintain continual relations with their home country, which are increasingly also facilitated by modern technologies of transport and communication (247).[13]

While he does not focus on Mexican immigration in particular, Arjun Appadurai, another foremost thinker of diaspora, has similarly argued that it is the "images, scripts, models, and narratives that come through mass mediation . . . [which] make the difference between migration today and the past" (6). In *Modernity at Large,* Appadurai emphasizes the confluence of two globalizing forces—improved travel and mass mediation—which now allow a larger number of people to begin to even imagine (and often act upon) the possibility of travel and displacement. Appadurai suggests that these cultural forces of globalization enable immigrants to continuously move between the United States and their home countries and/or to form deterritoralized diasporas with the help of communication media.

Both theorists converge in their assumptions about the end product of contemporary migration. Clifford argues that since constituents of diasporas are unevenly assimilated to the U.S. nation-state, they forge post-national forms of attachments that oppose hegemonic notions of citizenship. The resulting allegiances and connections of immigrants to their homeland or to a dispersed community elsewhere articulate alternate public spheres, which are made up of distinct political cultures of resistance (Clifford 265). In a very similar move, Appadurai theorizes the emergence of "diasporic public spheres" that undermine nationalist ideologies and repressive state mechanisms. As they become aligned with other progressive forces of de-nationalization (like communication media), such diasporic transnational formations gesture toward the emergence of more globally and transnationally organized entities as alternatives to some of the political failures of the nation-state.

The Border Patrol State

It is true that new and improved types of travel and communication have encouraged a certain growth in cross-border flows and have helped produce the contemporary figure of the transmigrant, for whom continuous and systematic border crossings of various kinds have become central. Theories foregrounding the anti-nationalist force of diasporas have, however, not been especially attentive to changing juridical-political constructions of diasporic subjects or to the increasing diversification among contemporary immigrants. This work has not adequately addressed why the much celebrated opening of national borders has benefited only certain kinds of (potential) border crossers, at the same time that it has enabled the unbridled flow of capital and goods under the neo-liberal politics of border-free economics.

Contemporary U.S. provisions for entry, codified in NAFTA and new immigration laws, afford certain people with economic (and, in part, cultural) capital—such as high-tech professionals and entrepreneurs—increased access to new modes of globalization that include open borders. At the same time, other forms of potential immigration have become further curtailed by stricter legislation and tightened border controls. Passed in 1991, the original FTA between Canada and the United States became the North American Trade Agreement (NAFTA) in 1994 after the inclusion of Mexico. Specifying the reduction or eradication of tariffs charged on traded goods between Canada, the United States, and Mexico, the agreement alleviates the flow of goods and services.[14] But NAFTA makes virtually no provisions for the free movement of people across the two international borders. It neither mentions issues of migration nor considers the possibility of new forms of hemispheric citizenship. Instead, the agreement only improves conditions of entry (into the United States and Canada) for business people and highly specialized professionals with competency in computer and media technology (Jamieson et al. 250).

NAFTA thus represents the transnationalization of general tendencies within U.S. immigration law to increasingly differentiate among various segments of migrants by a list of criteria that make immigrants either economically or educationally "useful" to the United States. After the 1965 abolition of national or ethnic criteria in immigration law, the introduction of skill-based and economic criteria encouraged a surge in elite transnationalism.[15] At the same time, however, other reforms further restricted the entry of non-elite immigrants by raising the income threshold required to sponsor relatives and by tightening eligibility for political asylum—one of the only available legal recourse for those arriving in the United States without documentation (Jonas 75).[16]

Similar to developments in other parts of the world, U.S. conditions for entry are thus becoming increasingly formalized into citizenship criteria that divide contemporary immigrants into several, hierarchically defined groups.[17] In comparison to nineteenth century U.S. law which openly excluded immigrants on the basis of ethnic and national origin by barring entire groups of people, today's immigration policies (with certain exceptions)[18] stratify migrants according to class, educational, and social background as well as the kind of classed position into which they will eventually be inserted in the United States. As Roger Rouse has argued, the purpose of these highly differentiated boundary regulations is not so much to keep immigrants from entering the United States, as official rhetoric will have us believe, but to shape the conditions under which they enter and eventually live (Rouse "Where").

Those who share a common national origin or a common past in discriminatory immigration legislation may thus no longer have a common future in the contemporary United States. Some immigrants with economic and/or cultural capital are immediately able to benefit from the privileges associated with U.S. residency or citizenship, while the access of others to a diminishing set of resources has become increasingly restricted. According to Aihwa Ong, powerful elements of the (often ethnically Chinese) Hong Kong and Singaporean capitalist class, for example, now generally bypass the conventional generational model of ethnic succession. Instead of working their way up through the second generation, these affluent Asian immigrants tend to move directly into the upper reaches of U.S. American society (Ong 174). This is not to say that even elite immigrants are not also faced with a certain lack of cultural capital or that they no longer encounter racism in the United States, but that contemporary migrant groups from Third World countries, even those from the same region or of the same ethnic background, are internally stratified along what often resembles class lines.

As I have argued earlier, the U.S.-Mexico border has not always been militarized, even though discourses about this region now often make it seem that way. Despite the erection of some fences in 1924 to enforce new restrictive quotas for European and Asian immigrants at the U.S.-Mexico border, until the late 1970s and 1980s, both U.S. land borders remained relatively open. In fact, a 1970 immigration bill submitted to Congress by Peter W. Rodino, Jr (D-NJ), argued for the exemption of Canada and Mexico from immigration quota in "recognition of the common undefended borders we share with these countries" (qtd. in Murata 10). Only in the late 1970s, after more than five decades of supervision by the Border Patrol, were long stretches of the southwestern border *systematically* re-enforced with chain link fences (Dunn 38). Gloria Anzaldúa has vividly described this so-called tortilla curtain in her account of *Borderlands/La Frontera* by portraying it as a "steel curtain / chainlink fence crowned with rolled barbed wire" (2).

Since the early 1990s, however, the border has become even further militarized. In her novel, *Tropic of Orange,* Karen Tei Yamashita describes the border as being enforced by "seismic sensors and thermal imaging, / . . . steel structures, barbed wire, infrared binoculars, / [and] INS detention centers" (199). Within the span of the decade that separated the publication of Anzaldúa's and Yamashita's texts, many of the chain link fences had not simply been maintained or expanded but *replaced* by several varieties of solid steel walls, and the Border Patrol had been upgraded to the largest U.S. federal law enforcement agency. Both developments have made the southwestern border harder for undocumented people to cross than ever before in U.S. history. Often simply belittled as "fences" (a term that invokes the older image of the "tortilla curtain"), many of these structures were actually welded from the steel of corrugated military aircraft landing mats or other military material (Dunn 66). In their solidity, these obstacles resemble the former long-time division of East and West—the Berlin Wall—which U.S.

foreign policy perceived as a major obstacle to the worldwide spread of freedom and democracy.[19]

Borderlands/La Frontera

The reinforcement of U.S. land borders and its relationship to undocumented (or improperly documented) immigrants have, at best, become the backdrop of a theoretical emphasis on travel, routes, and displacement within work on borders. In fact, border studies is characterized by many of the same assumptions that have become central to diaspora theory. While analyses of borders have simply not yet dealt with the U.S.-Canada frontier, the qualitative change in the nature of U.S.-Mexico border militarization and its effects on the increasing segmentation of immigration has largely become obscured by an overwhelming focus on de-nationalization. This perspective tends to foreground the continuity of border militarization since the creation of Mexican frontier rather than emphasize its qualitative change.

Border intellectuals have developed a general view of the border as a site of fluidity and porosity that is bridged by transnational ethnoscapes.[20] Their approach to borders is centrally rooted in Chicana/o civil rights struggles and ideologies of cultural nationalism (*Chicanismo*). *Chicanismo* originally focused on reclaiming the U.S. Southwest from its long history of Mexican and Mexican American racialization. Vividly summarized in the concept of Aztlán, *Chicanismo* reconfigured the Southwestern geography into an imaginary site for the creation of an independent Chicana/o nation. The notion of Aztlán helped to forge a distinct Chicano identity out of Mesoamerican, pre-Cortésian, and pre-Columbian myths that was set in opposition to official U.S. historiography and identified contemporary forms of oppression with the loss of Mexican land to the United State in 1848. At the same time, however, *Chicanismo* also contained the seeds for a hemispheric perspective as articulated in its founding manifesto, "El Plan Espiritual de Aztlán," in the notion that "we do not recognize capricious frontiers on the Bronze Continent" (1). Marxist-inspired sections of the Chicano rights movement, in particular, developed a pan-American view that joined Chicana/os with working-class Mexicans and thus already espoused what we would today term a diasporic perspective.

In the late 1980s, when parts of the U.S.-Mexico border had become reinforced by chain link fences designed to "control" Mexican immigration, these cultural nationalist perspectives on the border seem to undergo some change. As manifested in Gloria Anzaldúa's 1987 *Borderlands/La Frontera,* the utopian notion of a future homeland Aztlán *without* borders became transfigured into the concept of borderlands as a location where several cultures have historically fused and mixed *in spite of* the history of the international border. Anzaldúa's view suggests that at a time when the border had become militarized with fences, utopian demands for the *abolition* of the U.S.-Mexico border had shifted to a certain *acceptance* of the realities of a reinforced international frontier.

Anzaldúa's notion of cultural fusion also imaginatively projects an alternative future for this site. Her view of the "1,950 mile-long open wound [as] divid[ing] a *pueblo,* a culture" (Anzaldúa 2) thus supplements the diasporic notion of Chicana/o identity as the historical product of a shifting international border (that had already been somewhat articulated in the concept of Aztlán) with her faith in the metaphorical reclamation of the U.S. Southwest through the continuous cross-border movement of (largely undocumented) Mexican immigrants despite reinforced borders. In this way, immigration has become reconfigured as a way to imaginatively take back the territory of the Southwest and to erode a border that is dividing a transnational people.

Similar views on Mexican immigration continue to inform border studies' perspective on contemporary contradictions between border-free economics and border militarization. Within border studies scholarship, these conflicts tend to be framed as struggles between the de-nationalizing tendencies of globalization and obsolete (stable and "rooted") forms of domination by the nation-state. Theories of continued undocumented immigration (despite border enforcements) emphasize individual immigrant agency, transnational immigrant networks, and the ongoing need of U.S. domestic businesses for migrant labor as major factors for immigration.[21] Border militarization, on the other hand, has become almost exclusively identified with state nationalism, in particular, the U.S. nation-state's outdated efforts to retain control over the racialized boundaries of U.S. American citizenship. The further hardening of borders is perceived as the state's attempt to (re-)gain control over the origins of labor migrancy and/or to "discipline" undocumented immigrants into accepting low wages, substandard working conditions, and exclusion from other benefits of citizenship.[22] While theories of contemporary immigration thus identify today's transmigrants with forces of de-nationalization and anti-statism, they continue to attribute border militarization and exclusionary laws to the nation-state.

Matters of State and Capital

This view also accepts a relatively stable relationship between the state and U.S. domestic businesses in allowing (and even, in a sense, creating) undocumented immigration by not completely enforcing existing exclusivist legislation. But the state's response to the continual domestic need for labor in the United States (in connection with other "pull" factors in the sending countries) alone cannot adequately explain why today's undocumented immigrant groups from Mexico and mainland China are arriving simultaneously and in much larger numbers than at the turn of the twentieth century.

Sociologist Saskia Sassen has suggested that immigration today has become primarily market- rather than state-driven. In *Globalization and Its Discontents,* she has identified the transnationalization of U.S. investment as the most decisive factor for transforming out-migration from specific countries into large-scale immigration. With regard to Mexico, intensified U.S. investment in labor extensive-*maquila* production, beginning with the Border Industrialization Program, has acted as a catalyst for massive immigration into the United States in the 1980s. Contrary to official expectations that NAFTA would reduce the number of *indocumentados* by providing more job opportunities in Mexico, production facilities have attracted migrants from Mexico's heartland. These migrants come to the United States when they realize that the only difference between a job in a *maquiladora* and a comparable job in the United States lies in the immense wage gap between the two. Given the demand for low-wage labor in the United States, for many migrant workers and a growing percentage of Mexicans in general, immigration thus becomes a very real possibility (Sassen 33).

Similar to post-NAFTA developments in Mexico, Chinese migration has been on the rise since the 1980s because of China's opening to international trade and the corresponding influx of foreign capital. U.S. and other foreign investment in export-oriented sectors have encouraged the establishment of special economic zones along the Chinese coast, where, as in *maquiladoras,* many young women are employed. The permanent granting of U.S. most favored nation status and the initiation of closer economic cooperation with Western European countries are further indications that China's development will probably come to more closely resemble that of the U.S.-Mexico borderlands. Thus, at the same time that foreign investment is creating economic development and jobs, it is also beginning to transform China into one of the most important suppliers of immigration to the United States and other countries. Current predictions forecast that immigration from all of Asia will approach the magnitude of out-migration from Mexico within the next five to ten years (Goldstone 66).

Let us consider, then, that undocumented immigration to the United States continues in spite of border militarization not only because of individual immigrant agency or diasporic kinship networks, but also because of the U.S. nation-state's contradictory enforcement practices. The paradoxes of regulation reflect the state's need to accommodate the (sometimes opposed) interests of both domestic and transnational businesses. On the one hand, the U.S. state shores up its borders against immigrants of certain non-elite backgrounds from countries where corporations depend upon a steady supply of cheap labor. Border militarization thus not only disciplines undocumented migrants into accepting considerably reduced benefits of residency and citizenship in the United States as Rouse has argued, but it may also encourage certain kinds of migrants from non-privileged backgrounds to stay in their home country and work there for transnational companies, thus reinforcing one of the fundamental principles of globalization—that labor stays local.

Just as early twentieth-century U.S.-Mexico border policing compensated for the lack of binational diplomacy with Mexico in enforcing Chinese exclusion, today's build-up at U.S. boundaries may, in fact, be taking over some of the functions of immigration legislation, which is no longer able or willing to openly target certain groups of people based solely on ethnic and racial characteristics. In other words, even though the United States has a long history of blatantly exclusionist and racist immigration restrictions against Mexicans and Chinese (and other Asians), only the context of the Border Industrialization Program and intensified global economic integration in the 1990s have created conditions for large-scale U.S. border militarization (characterized first by chain link fences and then steel walls) that significantly differs from earlier forms of border enforcements.

On the other hand, the U.S. state also continually provides enough (loop)holes in steel walls and immigration regulations to allow a certain amount of non-elite migrant influx to meet the labor needs of domestic businesses, especially those in border areas and metropolitan centers near U.S. borders. Thus, the state adopts a variety of policies that anticipate and (to an extent encourage) the influx of (a limited) amount of undocumented labor across newly militarized national borders, such as the deportation of immigrants to nearby sites from which they can easily cross again and the termination of INS raids on undocumented working immigrants. State sovereignty takes the form of military enforcement in the Southwest, which Leslie Silko has memorably termed "the border patrol state," and increasingly also in U.S.-Canada border areas.

While twentieth century undocumented immigration predominantly affected the southwestern U.S. border, the nature of the Canadian frontier today is also beginning to change. Although it will probably never become militarized with walls or fences (in part also because of the northern border's particular geology), since the 1990s the U.S.-Canada border has witnessed the progressive tightening of immigration legislation and border controls for the undocumented border crosser. The 1997 rule of "expedited removal," for example, accords Border Patrol agents at both U.S. land borders the power to bar non-U.S. citizens from entering for a period of up to five years if the authenticity of their credentials is in doubt. The new provision generally offers no legal recourse for appeals of the authorities' decision. According to recent U.S. newspaper reports, it is predominantly Mexican and Canadian nationals trying to cross either U.S. border who have been affected by this rule (DePalma 1 and 30).

In addition, the inclusion of Section 110 into the 1996 Immigrant Responsibility and Illegal Immigration Reform Act represented another effort to filter out unwanted immigrants at both borders. This provision requires the completion of registration entry and exit forms, and the payment of fees by all non-U.S. citizens crossing the borders with Mexico and Canada. Pushed as part of Newt Gingrich's Contract with America, it was designed as a conservative response to the alleged threat of illegal immigration based on overstayed visas. After massive protests by Canadian business groups and U.S. politicians from border states, Section 110 was eventually repealed in 2001. At the same time that restrictions against undocumented immigration are being put in place, however, it has become increasingly easier for certain high-tech workers and other highly educated Canadian residents to obtain U.S. work and resident permits under NAFTA regulations right at actual U.S. ports of entry.

It was thus the sudden discovery that Chinese and other undocumented immigrants are using Canada as a passageway into the United States rather than the generally increased flow of people across the U.S.-Canada border that first prompted calls for the reinforcement of the border with the "good neighbor to the north." Ever since the U.S. Border Patrol began apprehending dramatically higher numbers of undocumented immigrants along the northern border in the 1990s, U.S. officials have requested that Canada bring its immigration laws in line with those of the United States (P. Smith 5). The head of the Senate subcommittee on immigration, Senator Spencer Abraham of Michigan, asked in January 2000 that more U.S. Border Patrol agents be stationed along the Canadian frontier because, as he puts it, the "Southern border isn't the only place people can cross into the country and do so with the intent to commit crimes" ("Hiring" A3).

While demands for border enforcement were originally articulated within an anti-immigration framework, they have now reemerged in the guise of anti-terrorist measures. Whereas, in the mid-1990s, Congress still denied requests for 1,000 more Border Patrol agents to be stationed specifically at the Canadian not the Mexican frontier ("Law Enforcement 42), it just passed an anti-terrorist bill that asks for the tripling of the number of federal agents there (Turner B5).[23] The bill also revives the main idea behind Section 110 as it provides for the implementation of a "check-in, check-out" system for U.S. borders that would require visitors to have their names recorded every time they enter and leave the country. The Immigration and Naturalization Service plans to put the system in place at airports and seaports by 2003 and at the 50 largest land entry points by 2004 (Toulin n. pag.). The two U.S.

boundaries are thus becoming more rigid than ever to border crossings by undocumented immigrants from China and Mexico at a time when the character of U.S. frontiers is being transfigured by increasingly "borderless" trade, whose regulations, in turn, tend to benefit other immigrants and diasporic people.

Immigrant Acts[24]

Undocumented immigrants from Mexico and China respond to these contradictory demands placed on them by developing flexible strategies of dislocation. As border and diaspora theorists have suggested, some of these strategies entail the continual movement between the United States and countries of origin or the formation of deterritoralized diasporas via communications media. Aihwa Ong has, however, recently proposed a more complex account of such forms of "flexible citizenship." She argues that these cultural expressions should not only be viewed as means of contesting hegemonic notions of U.S. citizenship, but also as responses to contradictory demands in the labor markets of the United States and the immigrants' home country (Ong 2).

While immigrants' continual transnational movement may encourage them to escape their fixation into homogeneous notions of citizenship once in the United States, unauthorized border crossings practiced largely by non-elite immigrants have also remained fundamentally shaped by market regulations as well as by increasing violence at geopolitical borders. The escalation of border policing has helped to transform the once relatively simple act of border crossing into a more complex system of illegal practices that have become associated with sophisticated forms of smuggling. Undocumented immigrants today not only rely on highly organized networks of smugglers, but they also enter the United States in different areas. Just as Mexican immigrants have largely shifted their entry points from fortified metropolitan areas (e.g., San Diego and El Paso) to the dangerous deserts, Chinese smugglers have expanded their operations from the Mexican to the Canadian border. As the 1998 discovery of the "largest global alien smuggling ring" between the United States and Canada demonstrates ("Huge" A1), by the late 1990s international syndicates had successfully transformed the border with Canada into another major transit corridor linking China to the United States.

Between 1996 and 1998, a Chinese-controlled operation brought close to 4,000 undocumented immigrants, predominantly from the Chinese province of Fuzhou, across the U.S.-Canada border into New York City's Chinatown. Similar to the cooperation between Mexican and Chinese smugglers at the U.S.-Mexico border, the Chinese-controlled organization "sub-contracted" the actual crossing of the U.S.-Canada border to residents of the transnational Mohawk reservation of Akwesasne. Located on the St. Lawrence Seaway about 95 km southwest of Montreal, the reservation straddles New York, Ontario, and Quebec. The fact that residents claim exemption from several state, provincial, or federal U.S. and Canadian laws (including taxation on traded good and impediments to free travel), coupled with complex and overlapping claims to jurisdiction of this territory by several state, governmental, and tribal organizations makes border enforcement there more difficult.[25] As a result, the route across the Mohawk reserve has today become one of the most important U.S. gateways for undocumented Chinese. Similarly, since the fall of 1999, at least 100 Chinese have been apprehended every month while trying to cross the U.S.-Canada border via Ontario's Walpole Island First Nation reservation. The Canadian Indian reserve is separated from Michigan's Harsen's Island only by a narrow channel off the St. Clair River. When the river freezes, smugglers usually turn to the Ambassador Bridge—located between Detroit, Michigan, and Windsor, Ontario—and to the train tunnel connecting Port Huron, Michigan, and Sarnia, Ontario ("Island" n. pag.).

In choosing these kinds of transnational locations, some of which had already served as important entry points for Chinese immigrants at the turn of the twentieth century, individual immigrants and smuggling networks prove that they are able to successfully manipulate contested sites of transnationalism, be it reservation land or national differences in immigration legislation. Just as Chinese immigrants in the 1880s crossed into the United States from the territory of a country that, at that time, had not yet completely barred them under exclusionist laws, today many Chinese use Canada as a way station where they first apply for asylum status and then continue their journeys as undocumented immigrants into the United States. In comparison to U.S. provisions for entry, contemporary Canadian immigration law has remained far less exclusionary of non-elite immigrants.[26] While the largest U.S. immigrant groups have to cross land borders clandestinely because criteria for political asylum have become progressively more restricted, Canada still allows the majority of its new arrivals from Southeast Asia, the Middle East, and Africa to claim refugee status, whose requirements have been left relatively open. In contrast to the United States, which now largely implements the mandatory detention of refugees, Canada applies so-called balance-of-credibility tests that not only accord refugees claimants the benefit of doubt when trying to determine their eligibility, but that also guarantee Canadian constitutional rights to refugees and entering immigrants. Immigrants can apply for asylum upon arrival in Canada and their work permits are quickly approved while they wait for the resolution of their cases in a process that usually takes up to three years. Some later continue on to the United States.

In the aftermath of the September 11, 2001 attacks, however, the Canadian government has already buckled under sustained U.S. pressure and begun to tighten its refugee law and border enforcement policies. In October 2001, the Minister of Canada's Citizenship and Immigration announced the introduction of a new permanent resident card (the "Maple Leaf Card"), the implementation of tighter security screening for asylum seekers, the extension of detention policies, and the acceleration of deportations for immigrants as well as the hiring of 300 more border guards. These guards will be added to the nearly 2,400 customs officials and 560 immigration officers already employed at Canada's ports of entry (McNeil B3).

Canada is currently also under pressure to bring its immigration and refugee laws even more closely in line with U.S. regulations to comply with plans for the creation of a common "North American Security Perimeter." When the proposal was first articulated in summer 2001, it simply called for the tightening of screening measures at all external North American entry points and the simultaneous loosening of control at U.S. borders. Since September 11, however, it has been made clear to the Canadian government that the creation of such a continental security zone would require the "harmonization" of immigration policies, especially refugee claimant programs, visitor visa policies, and border policing between Canada and the United States (and potentially Mexico, as Vicente Fox has signaled his interest in participating in discussions about the Perimeter concept). There is talk about a potential cross-border refugee agreement, allowing the return of claimants who come to Canada via the United States, and about a "visa convergence" policy, under which Canada and the United States would assemble a list of countries whose citizens would require visas to North America. The United States has already signaled that Canadian failure to accept such an approach would mean a dramatic increase in security at the U.S.-Canada border with devastating consequences for cross-border trade, tourism, and Canada's ability to attract foreign investment. If history is any indication, the planned convergence of the two countries' border policies may once again simply end up re-directing immigration flows to other geographies.

I hope that my analysis of various border crossers' responses to the contradictory conditions of border enforcement and border-free economics has opened up for critique the assumptions underlying the current focus on the de-nationalizing effects of diasporic immigrant formations. This prevailing emphasis runs the danger of reducing complex global developments—including the formalization of global inequities—to questions of restrictive state governmentality by particular First World nations like the United States. It also simply acknowledges the spread of certain cultural forms that have remained organized around (and thus somewhat re-territorialize) older conceptualizations of the nation as bounded by a common culture, language, and/or descent.

While cultural, linguistic, and ethnic ties remain important for theories of social action, we also need to make various groups of borderlanders visible to one another by showing how similarities in their positions are mediated by the joint efforts of nation-states and corporations to accommodate processes of globalization. A rethinking of theories of affiliation would allow different understandings of political identities and social activism that could, for example, center on cross-ethnic and cross-diasporic struggles for immigrant rights, which are currently being further eroded in new anti-terrorism legislation. As evidenced in recent mass protests against meetings of the World Bank, the World Trade Organization, and against planners of NAFTA's expansion into the Free Trade Area of the Americas (FTAA)—all instruments and proponents of economic de-nationalization—it appears that such "unlikely coalitions"[27] that cut across and, in previously unimaginable ways, unify a wide variety of groups around common political goals may become the most effective tool in establishing alternative forms of grassroots transnationalism and globalization.

Notes

1. I would like to thank the members of the Fall 2000 Dartmouth Humanities Institute, "Los Angeles/La Frontera/Mexico City," for their careful reading of an earlier draft of this essay and for their excellent advice on revisions. My special thanks to Kenton T. Wilkinson, Emiliano Corral, Santiago Vaquera-Vásquez, Boris Muñoz, Silvia Spitta, Mark Williams, Ulises Juan Zevallos-Aguilar, Francine A'Ness, and Maarten Van Delden.
2. At the moment, the United States requires visas of visitors from all but 29 countries, whereas Canada exempts 58 countries (Clark A6).
3. Developments at the northern frontier have not yet become a focal point of the burgeoning interdisciplinary scholarship on borders. This work in anthropology, geography, political science, history, and sociology has recently expanded its initial focus on the U.S.-Mexico border—the birthplace of border studies—to other international boundaries in Europe, Africa, and Asia. Internationally oriented comparative work on borders foregrounds similarities of cultural syncretism, cross-border migration, border shopping, and illegality. For an excellent overview of this scholarship, see Hastings Donnan and Thomas M. Wilson. In comparison, cultural studies work on borders has largely retained its focus on the U.S.-Mexico frontier.
4. U.S. Americans tend to overlook (and, as manifested in popular cultural representations, often ridicule) Canadian attempts at stressing *national* (i.e., cultural and, to an extent, linguistic) differences from U.S. Americans. While U.S. Americans thus downplay the significance of their frontier with Canada, to Canadians—over 60 percent of whom live within 100 miles of

the U.S. border—this site has remained very important for issues of national self-definition. Not surprisingly, much of Canadian literature, which has become an object of academic study in the 1960s, incorporates and reflects on borders.

5. As it locates oppositional agency in the Chicana/o (and Mexican) border subject, border work has, from its inception, also opened up the possibility for the dis-identification of the national frontier from its actual geography to signify a certain "portable" *ethno-cultural* identity. For a critical view of the way in which Chicana/o studies has claimed the U.S.-Mexico border, see Scott Michaelsen and David E. Johnson's introduction to *Border Theory.* The equation of "border matters" with analyses of a particular racialized group has, however, not been limited to American and cultural studies. In their overview of work on international borders, Hastings Donnan and Thomas M. Wilson have similarly argued that studies of the U.S.-Mexican border in anthropology, history, sociology, and political geography still focus most prominently on Mexican and Mexican American life (14).
6. Roger Rouse, for example, has interrogated the emergence of a "transnational migrant circuit" as a formation that is becoming increasingly important for the organization of Mexican migration to and from the United States. Rouse has investigated the continuous movement from a rural *municipio* of Aguililla in the state of Michoacán to Redwood City, California, which is creating a "transnational space" or a "kind of border zone" (Rouse 15). In response to new forms of globalization and to declining chances for upward mobility in the United States, migrants from Aguililla maintain active relationships with Mexican friends and relatives, sustain two distinct ways of life, and often send their children back to Mexico to endow them with bilingual and bicultural skills, which are necessary to function on both sides of the border. Immigrants from Central America also attach themselves to the transnational space in Redwood City, and should they need to leave the United States, they can call on social ties in Aguililla rather than having to return to their own country. Rouse's scholarship has been employed to formalize the currently dominant view of Mexican immigrants and their descendants as constituents of a Mexican or increasingly a "Hispanic" diaspora that is not entirely unproblematic. Sociologist William Safran, for example, has criticized the application of diaspora to Mexican Americans. Since it no longer denotes the violent dispersal from a homeland to multiple locations, in this reconfiguration the term diaspora names a group that has not been exiled, expatriated, or physically removed from an origin but whose contemporary displacement from the homeland is rather the result of demographic changes. In addition, Safran does not consider Mexican Americans to be part of a diaspora because they are assimilating to the dominant culture and do not cultivate a myth of return. He argues that their homeland cannot be easily idealized since it is geographically near (Safran 90).
7. Here I am using the title of sociologist Peter Kwong's book on recent undocumented Chinese immigration to the United States.

8. Whereas the United States and Canada began regulating their respective immigration flows in the nineteenth century, the Mexican government has taken no steps to curb undocumented immigration. In fact, Mexico's recent policies of economic growth without significant gains in employment help promote out-migration. Especially since the 1970s, Mexico has become increasingly reliant on exporting its labor surplus. Entire sending communities have become economically dependent on remittances from migrant workers that provide much needed foreign exchange.
9. According to Erika Lee, annual reports from the Bureau of Immigration indicate that between 70 to 200 Chinese immigrants were apprehended annually at the Mexican border between 1882 and 1901 (Lee "Crossing Borders"). As a result of the 1921 Quota Law, which placed limits on the number of immigrants from each European country, many Europeans who had formerly been able to enter the country legally through various ports, now tried to come to the United States "illegally" across the U.S.-Mexico border. Whereas the European undocumented border crossers were usually seen as victims of border migration schemes, however, the much smaller number of "illegal" Chinese was systemically racialized and viewed as threats to the U.S. nation. As customs inspector Clifford Perkins reports in his memoirs, except for an organized group bringing Greeks into the United States and for some Italian immigrants who were suspected of belonging to the Mafia, the smuggling of European aliens across the border appeared to be a relatively unplanned operation (Perkins 79).
10. Out-migration from the United States to China also included U.S.-born Chinese. These flows of people responded to the repressive treatment that kept Chinese (Americans) residentially segregated in ethnic enclaves and excluded them from access to many jobs. According to Gloria H. Chun, about 20 percent of American-born Chinese returned to China in the 1930s (Chun 174).
11. Until the late 1980s, Mexican undocumented workers tended to circulate between Mexico and the United States because their illegal status constituted a deterrent to more permanent settlement. Between 1965 and 1986, only 16 percent of the 27.9 million undocumented Mexicans who entered the United States actually stayed; the rest remained there only temporarily (Durand et al. 535).
12. The repeal of various Chinese Exclusion laws, allowing Chinese naturalization and the annual immigration of 103 Chinese, had only a small measurable effect on the actual number of immigrants (Yu 76). For example, while in the 1930s about 200 Chinese were admitted into the United States, in 1940, the number of Chinese rose to 920 (Chun 177).
13. In his cross-cultural work on travel and diaspora, Clifford recognizes some degree of distinction among various movements by differentiating between "travelling-in-dwelling" and "dwelling-in-travelling." In contrast, I insist on recognizing dissimilarities between the travels of immigrants as a result of legislation that accommodates the emergence of an increasingly international infrastructure and that creates enormous juridical-political distinctions between

the undocumented and other migrants, such as entrepreneurs and business-people.

14. Since NAFTA, trade between Canada and the United States has increased to the extent that the Detroit-Windsor border crossing has been officially titled the "NAFTA Superhighway," because 45 percent of all the U.S.-Canada trade moves through this port of entry ("Law Enforcement Problems" 56). Mexico, in turn, had by 1988 surpassed Japan as America's second largest trading partner behind Canada. Similarly, NAFTA has encouraged the movement of a larger amount of U.S. capital from the United States and other localities into labor-intensive, export-processing zones of *maquiladora* (assembly) production in Mexican border towns ("Mexican Makeover" 51). Drawing on pre-existing binational agreements between the United States and Mexico, such as the Border Industrialization Program, NAFTA has both formalized and accelerated the re-location of export-processing production facilities into less developed countries, while crucial research capacities and the majority of consumers have remained firmly anchored in First World metropolises. Such economic restructuring has largely been characterized by "post-Fordist" regimes of accumulation, i.e., greater flexibility in production, in the relationship between core firms and suppliers, and in the organization of labor processes (Kiely 98).
15. The 1986 Immigration Reform and Control Act (IRCA) introduced distinctions between workers who were already living in the United States and new "illegal" arrivals. While it allocated more funds to the Border Control's mission to constrict immigrant flows, it also offered amnesty and regularization to undocumented farm workers and long-term settlers already in the country. IRCA transformed many of the immigrants who had previously been shuttling back and forth into more permanent settlers, but its employer sanctions also contributed to the worsening of working conditions and wages for undocumented labor (Durand et al. 518). Whereas before IRCA undocumented and documented migrants earned similar wages, with rates of pay being determined largely by education, English language ability, and U.S. experience, after the passage of the new immigration law, jobs began to be determined by a person's social contacts, and undocumented workers could be hired under worsened conditions (Durand et al. 528).
16. I am thinking here of the 1996 Illegal Immigration and Immigrant Responsibility Act (IIIRA), the 1996 Personal Responsibility and Work Opportunity Reconciliation Act (Welfare Reform Act), and the 1996 Anti-Terrorism Bill. The Welfare Reform Act, in particular, signals what may be the beginning of a move toward more binary distinctions on the basis of citizenship as it bars non-citizens (both undocumented and legal immigrants) from receiving the benefits of federal and state social programs.
17. Consider, for example, a recent Swiss move for a referendum that would institute an upper limit for immigration, which, if enacted, would not affect special provisions for professionals in the information technology industries.

18. As an ideological holdover from the Cold War era, one of the exceptions to current U.S. immigration policy is practiced toward citizens of Cuba.
19. There is disagreement about the effects of border militarization on undocumented immigrants. While scholars like Wayne Cornelius insist that new obstacles will not deter immigrants from crossing, others like Michael Smith and Bernadette Tarallo have pointed to the diversion of border crossing activities into other, less enforced (and mostly more dangerous) areas along the border.
20. Gloria Anzaldúa, for example, begins her account of borderlands with a very memorable depiction of U.S.-Mexico border enforcements. But already on the second page of her book she turns to a description of the "Tortilla Curtain" fence as "running down the length of my body, /staking rods in my flesh, / splits me splits me / *me raja me raja* / This is my home / this thin edge of barbwire" (Anzaldúa 2–3). In this shift, Anzaldúa transforms the fortified border into a metaphor that signals her own divided cultural identity.
21. See, for example, the work of Singer et al., Rodríguez, and Behdad "Nationalism and Immigration."
22. See, for example, Rodríguez and Behdad "INS and Outs."
23. This request responds to the fact that, compared to the about 7,000 agents currently stationed at the Mexican frontier, in 1998 less than 300 Border Patrol agents controlled the Canadian border ("Law Enforcement" 43). The difference constitutes further proof that, throughout the 1990s, U.S. immigration policy primarily focused on *preventing* undocumented immigration at the southern U.S. border and expended less effort on *enforcing* existing legislation concerning visa policies that tends to also affect the northern U.S. border. Despite the passage of the 2001 anti-terrorism bill, it appears, however, that Congress will only appropriate enough money to add about 800 customs agents. A new appropriations bill only asks that priority will be given to the northern border, but contains no provisions to prevent INS bureaucracy from continuing its policy of shifting agents to other locations, especially in the U.S. Southwest (Turner B5).
24. This is the title of cultural theorist Lisa Lowe's pioneering work on the Asian diaspora, in which she theorizes similarities between the exploitation of Asians by U.S.-owned companies and Asian Americans in the United States.
25. For a more in-depth account of some of the issues affecting transnational borderlands tribes, see Donald Grinde's essay in this volume. Ruth Jamieson describes the situation at the Mohawk Territory of Akwesnasne in more detail in her article "'Contested Jurisdiction Border Communities' and Cross-Border Crime." She points out that the reservation is under the jurisdiction of three sub-national systems—the provinces of Ontario, Quebec, and New York State—so that policing has at various times involved the Royal Mounted Canadian Police, the Canadian Army, the FBI, the Ontario Provincial Police, the Sureté du Québec, the New York State Police, the Akwesasne Mohawk Police and the self-appointed Mohawk Sovereignty Security force. To complicate matters further, the St. Regis Mohawk Council, set

up by the New York State government in 1802, assumes jurisdiction over the American side of the reservation and the Mohawk Council of Akwesasne, set up by the Canadian government in 1899, exercises jurisdiction over the Canadian side, while the Mohawk National Council claims jurisdiction over the whole community (261).

26. As Barbara Driscoll has pointed out, differently from U.S. law, the notion of the refugee has remained fundamental to Canadian immigration legislation. Since 1962, Canada has received mainly three types of immigrants: family reunifiers, refugees, and those judged for suitability of entrance along the continuum of a skill-, language-, and capital-based point system. These differences are also reflected in the naming of the individual nation's institutions that are endowed with enforcement tasks, namely the Canadian Immigration and Refugee Board and the U.S. Immigration and Naturalization Service. Canada's refugee policy has had a significant impact on the very different ethnic composition of entrants as compared to the United States: Whereas in Canada the largest groups of refugees come from South East Asia, the Middle East, and Africa, in the United States the most numerous groups of immigrants are from Mexico and Latin America.
27. Here I am appropriating the title of a sub-section in Lisa Lowe's and David Lloyd's collection of essays *The Politics of Culture in the Shadow of Capital.*

Works Cited

Anzaldúa, Gloria. *Borderlands/La Frontera: The New Mestiza.* San Francisco, CA: Aunt Lute Books, 1987.

Appadurai, Arjun. *Modernity at Large: Cultural Dimensions of Globalization.* Minneapolis, MN: University of Minnesota Press, 1996.

Behdad, Ali. "INS and Outs: Producing Delinquency at the Border," *Aztlán* 23.1 (Spring 1998): 103–113.

———. "Nationalism and Immigration to the United States," *Diaspora* 6.2 (Fall 1997): 155–178.

"A Brief History of the Border Patrol," U.S. Border Patrol National Museum. undated: n. pag. Available at <http://www.flash.net/~prog2/bphistory.htm>. Accessed May 30, 1998.

Bridger, Chet. "Nearby Canadians Feeling U.S. Slump," *Buffalo News* (October 14, 2001): B11 and B12.

Chang, Gordon H. "Writing the History of Chinese Immigrants to America," *The South Atlantic Quarterly* 98.1–2 (Winter-Spring 1999): 135–142.

Chin, Ko-Lin. *Smuggled Chinese: Clandestine Immigration to the United States.* Philadelphia, PA: Temple University Press, 1999.

Chun, Gloria H. "'Go West . . . to China': Chinese American Identity in the 1930s," *Claiming America: Constructing Chinese American Identities During the Exclusion Era.* Eds. K. Scott Wong and Sucheng Chan. Philadelphia, PA: Temple University Press, 1998: 165–190.

Clark, Campbell. "Canada in Talks with U.S. on Pact Dealing with Refugees, Visitor Visas," *The Globe and Mail* Toronto (October 24, 2001): A17.

Clifford, James. *Routes: Travel and Translation in the Late Twentieth Century.* Cambridge, MA: Harvard University Press, 1997.

Cornelius, Wayne. "Impact of the 1986 U.S. Immigration Law on Emigration from Rural Mexican Sending Communities," *Population and Development Review* 15.4 (1989): 689–705.

DePalma, Anthony. "New Rules at U.S. Borders Provoke Criticism," *New York Times* (November 14, 1997): 1 and 30.

Donnan, Hastings and Thomas M. Wilson. *Borders: Frontiers of Identity, Nation, and State.* Oxford: Berg, 1999.

Dorning, Mike. "Visa Procedures Leave Country Vulnerable," *Buffalo News* (September 27, 2001): A1 and A5.

Driscoll, Barbara A. "Comparative Migration Issues," *Critical Sociology* 21.2 (1995): 67–74.

Dunn, Timothy J. *The Militarization of the U.S.-Mexico Border, 1978–1992: Low-Intensity Conflict Doctrine Comes Home.* Austin, TX: CMAS Books, 1996.

Durand, Jorge, Douglas S. Massey, and Emilio A. Parrado. "The New Era of Mexican Migration to the United States," *The Journal of American History* (September 1999): 518–536.

Gilroy, Paul. *The Black Atlantic: Modernity and Double Consciousness.* London: Verso, 1993.

Goldstone, Jack A. "A Tusnami on the Horizon? The Potential for International Migration from the People's Republic of China," *Human Smuggling: Chinese Migrant Trafficking and the Challenge to America's Immigration Tradition.* Ed. Paul J. Smith. Washington D.C.: Center for Strategic and International Studies, 1997: 48–75.

Gutiérrez, Carlos Gonzáles. "Fostering Identities: Mexico's Relations with Its Diaspora," *The Journal of American History* (September 1999): 545–567.

Gutiérrez, Ramón A. "Hispanic Diaspora and Chicano Identity in the United States," *The South Atlantic Quarterly* 98.1–2 (Winter/Spring 1999): 203–215.

"Hiring of Border Guards Falls Short," *The News Journal* Wilmington, DE (January 5, 2000): A3.

"Huge Alien-Smuggling Ring Used Canada's Refugee System," *Globe and Mail* Toronto (December 11, 1998): A1.

"Island Channel a Magnet to Illegal Immigration from Asia," *Detroit Free Press* (February 9, 2000). Available at <http://www.freep.com/news/statewire/sw6128_20000209.htm>. Accessed February 9, 2000.

Jamieson, Ruth. "'Contested Border Communities' and Cross-Border Crime—The Case of Akwesasne," *Crime, Law, and Social Change* 30 (1999): 259–272.

Jamieson, Ruth, Nigel South, and Ian Taylor. "Economic Liberalization and Cross-Border Crime: The North American Free Trade Area and Canada's Border with the U.S.A.," *International Journal of the Sociology of Law* 26.2 (1998): 245–272 and 26.3 (1998): 285–319.

Jonas, Susanne. "Rethinking Immigration Policy and Citizenship in the Americas: A Regional Framework," *Social Justice* 23.3 (Spring-Summer 1996): 68–85.

Kiely, Ray. "Globalization, Post-Fordism and the Contemporary Context of Development," *International Sociology* 13.1 (March 1998): 95–115.

Kwong, Peter. *Forbidden Workers: Illegal Chinese Immigrants and American Labor.* New York: The New Press, 1997.

LaDow, Beth. *The Medicine Line: Life and Death on a North American Borderland.* New York: Routledge, 2001.

"Law Enforcement Problems at the Border between the United States and Canada: Drug Smuggling, Illegal Immigration, and Terrorism," Hearing before the Subcommittee on Immigration and Claims. (April 14, 1999). Washington, D.C.: U.S. Government Documents, 1999.

Lee, Erika. "Crossing Borders and Race: Illegal Chinese Immigration During the Exclusion Era," Paper Presented at the American Studies Association Meeting, Montreal, Canada, October 1999.

———. *At America's Gates: Chinese Immigration During the Exclusion Era, 1882 - 1943.* Chapel Hill, NC: University of North Carolina Press, forthcoming 2002.

Lowe, Lisa. *Immigrant Acts: Asian American Cultural Politics.* Durham, NC: Duke University Press, 1996.

Lowe, Lisa and David Lloyd, eds. *The Politics of Culture in the Shadow of Capital.* Durham, NC: Duke University Press, 1997.

Massey, Douglas. "Foreword," *Smuggled Chinese: Clandestine Immigration to the United States.* Ko-Lin Chin. Philadelphia, PA: Temple University Press, 1999.

McIlwraith, Thomas F. "Transport in the Borderlands, 1763–1920," *Borderlands: Essays in Canadian-American Relations.* Ed. Robert Lecker. Toronto: ECW Press, 1991: 54–89.

McNeil, Harold. "New Strategy Announced to Bolster Border Security," *Buffalo News* (October 13, 2001): B3.

"Mexican Makeover," *Business Week* (December 21, 1998): 50–52.

Michaelson, Scott and David E. Johnson, eds. *Border Theory: The Limits of Cultural Politics.* Minneapolis, MN: University of Minnesota Press, 1997.

Murata, Katsuyuki. "The (Re)Shaping of Latino/Chicano Ethnicity through the Inclusion/Exclusion of Undocumented Immigrants: The Case of LULAC's Ethno-Politics," *American Studies International* 39.2 (June 2001): 4–33.

Ong, Aihwa. *Flexible Citizenship: The Cultural Logics of Transnationality.* Durham, NC: Duke University Press, 1999.

Perkins, Clifford Alan. *Border Patrol: With the U.S. Immigration Service on the Mexican Boundary 1910 - 54.* El Paso, TX: Texas Western Press, 1978.

Rodríguez, Néstor. "The Battle for the Border: Notes on Autonomous Migration: Transnational Communities, and the State," *Social Justice* 32.3 (1996): 21–37.

Rotella, Sebastian. *Twilight on the Line.* New York: W.W. Norton, 1998.

Rouse, Roger. "Where the Boundaries Lie: Mexican Migration in Transnational Times," Presentation at Dartmouth College as part of the Humanities Institute "Los Angeles/*la frontera*/Mexico City," November 8, 2000.

———. "Mexican Migration and the Social Space of Postmodernism," *Diaspora* 1.1 (Spring 1991): 8–23.

Safran, William. "Diasporas in Modern Societies: Myths of Homeland and Return," *Diaspora* 1.1 (Spring 1991): 83–99.

Sassen, Saskia. *Globalization and Its Discontents.* New York: New Press, 1998.

Saldívar, José David. *Border Matters.* Berkeley, CA: University of California Press, 1997.

Silko, Leslie Marmon. *Yellow Woman and a Beauty of the Spirit.* New York: Simon and Schuster, 1996.

Singer, Audrey and Douglas S. Massey. "The Social Process of Undocumented Border Crossing Among Mexican Migrants," *International Migration Review* 32.3 (Fall 1998): 561–592.

Smith, Michael Peter and Bernandette Trallo. "Preposition 187: Global Trend or Local Narrative? Explaining Anti-Immigrant Policies in California, Arizona, and Texas," *Journal of Urban and Regional Research* 19 (1995): 664–76.

Smith, Paul J. "Introduction," *Human Smuggling.* Washington, D.C.: Center for Strategic and International Studies, 1997: i-xvi.

Toulin, Alan. "U.S. Border Plan Forces all Visitors to Register," *National Post* (October 23, 2001). Available at http://www.nationalpost.com. Accessed October 28, 2001.

Turner, Douglas. "Washington Fails to Deliver Customs Agents," *Buffalo News* (November 18, 2001): B2 and B5.

United States Immigration and Naturalization Service. "Illegal Alien Resident Population," 1996: n. pag. Available at <http://www.ins.usdoj.gov>. Accessed May 30, 1998.

Walker, William. "Canada, U.S. Eye Scrapping Border," *The Star.* Toronto. (July 28, 2001): A1 and A25).

"Where Are You From, Originally?" *USA Today* (March 23, 2000): 1A.

Yamashita, Karen Tei. *Tropic of Orange.* Minneapolis, MN: Coffee House Press, 1997.

Yu, Renquiu. "'Exercise Your Sacred Rights': The Experience of New York's Chinese Laundrymen in Practicing Democracy," *Claiming America: Constructing Chinese American Identities During the Exclusion Era.* Eds. K. Scott Wong and Sucheng Chan. Philadelphia, PA: Temple University Press, 1998: 64–94.

Performing the Border

On Gender, Transnational Bodies, and Technology[1]

Ursula Biemann

Photo 4.1. Still from P*erforming the Border* by Ursula Biemann, video essay, 43 min, 1999. Photograph courtesy of the artist.

My video essay *Performing the Border* opens with a shot from inside a car moving through the Mexican desert near Ciudad Juárez. In the off, Bertha Jottar comments: "You need the crossing of bodies for the border to become real, otherwise you just have this discursive construction.

There is nothing natural about the border; it's a highly constructed place that gets reproduced through the crossing of people, because without the crossing there is no border, right? It's just an imaginary line, a river or it's just a wall."[2]

In this shot I was filming the woman driving the car and thus I, myself, inevitably became a part of the road narrative that unfolds as Bertha Jottar characterizes the U.S.-Mexican border as a highly performative place. It is a place that is constituted discursively through the representation of the two nations and materially through the installation of a transnational, corporate space in which different national discourses are both materialized and transcended. It is an ambivalent space at the fringes of two societies, remotely controlled by their core powers.

The Export Processing Zone

In artificial, post-urban industrial parks that stretch over large desert areas, U.S. corporations assemble their electronic equipment for the communications industry. Whereas the capital-intensive operations remain in the North, the labor-intensive operations are located south of the border. Within a short time the *maquiladoras*—the Golden Mills—have introduced a new technological culture of repetition, registration, and control into the desert cities. It is here that microelectronic components are being made for use in medical instruments, in information processing, in cyber-satellite systems, identification and simulation technologies, and in optical instruments for the aeronautic and military industry.

There is nothing natural about this transnational zone, and, looking at postmodern theory, it may even be that there's nothing *real* about it either. It is an entirely simulated place with simulated politics, a zone from which the idea of the public has been thoroughly eradicated. Housing, water, transportation, telephone wires, power supply, street lighting, sewage system, health care, childcare, and schooling have become the responsibility of the individual and, consequently, the site of spontaneous community initiatives. These ad hoc formations struggle to provide the most rudimentary social services. Any humanist claim is out of order in this sort of place. The post-human age, expressed in futuristic imagery, computer-generated by the hottest designer tanks in the North, is living out its dark side here on the border.

I don't mean to demonize technology. The way my own existence as a resident of the wealthy North has been transformed by the new media into a more connected, mobile, and accelerated lifestyle is certainly one of the reasons why I decided to go to this high-tech site of production and make it, if only temporarily, into my own production site of digital visualization. In this essay, it will become clear, however, that the kinds of subjectivities (like mine) that a new transnationalism brings forth in the North are radically different from the ones it produces in the South. The representations of transnational subjects that are inspired by global capitalism differ greatly. While discourses about residents of technology-consuming societies tend to efface their internal differences to establish a uniform transnationally mobile consumer, those on the producing end become even more over-determined and restricted by gendered, sexualized, racialized, and nationalized representations.

I recognize the growing need in cultural practice to locate questions of gender and other categories of identity, such as ethnicity and nationality, within the context of the wider transformations of the public sphere, particularly of urban reality. The question that interests me the most is how prevailing representations relate to the material reality of a specific site, that is, how the border as a metaphor for various kinds of marginalizations becomes materialized, not only in architectural and structural measures but also in corporate and social regulations of gender. I focus, therefore, on the circulation of female bodies in the transnational zone and on the regulation of gender relations in representation, in the public sphere, the entertainment and sex industry, and in the reproductive politics of the *maquila.*

A vital resource for this research comes from the rich cyber-feminist theoretical debate and art practice that has recently begun to explore the entanglement of the female body with technology and image production. I will draw on these contemporary feminist discussions to examine border identities and their multifaceted subversive potential. Further, I will propose a reading of the serial killings in Ciudad Juárez, which provides a glimpse of how urban politics, serial sexual violence, and technology converge in a dramatic way to reveal deeper layers of the psycho-social meaning of the border.

For a number of reasons, the assembly plants here originally attracted mainly young women into their labor force. When the first *maquiladoras* settled in the area in the mid-1960s, 95 percent of its labor force was female. In the last decade, the percentage of male workers has grown to about 40 percent in some areas, but in the micro electronic industry, the great majority of workers are still female. Every day hundreds of women arrive in Ciudad Juárez, which is across the Rio Grande River from El Paso, Texas. These women make up the majority of the population of the border town. They have created new living spaces, and they consume their own entertainment culture. Moreover, they have changed social structures and gender relations, and in doing so they are rewriting the texts of their bodies and of their society. These women are the ones who produce the instruments that enable the kind of cyberspace that affords millions of others north of the border the mobility and the freedom to consume. At the same time, the workers' own mobility remains confined to the outer limits of the "free zone" of post-fordist manufacturing. They are the newest members of transnationalism, but their citizenship in this formation functions on very different terms than that of the transnational consumer.

Communicating Borders

In the language of corporate offshore operations, the designations for the U.S.-Mexican border zone are very explicit. Companies either set up or close down shop wherever the conditions are optimal for them. Within this profit-driven framework, any facility or any person can be thought of in terms of dis-assembly and re-assembly. In fact, corporate terms used to describe the assembly process can be easily transferred onto the person doing the work. In the process, the worker becomes associated with language normally used to speak about machines, like speed, efficiency, and production numbers. Moreover, by translating human labor into robotics, the assembly process renders the worker's body inseparable from the machine she works at. Her body functions not only come to resemble those of the machine, but she is also rendered "post-human" in that the distinctions between the organic and the machine disappear. The body of the worker becomes "technologized" through the nature of assembly work and through a post-human terminology used to describe it. These processes fragment her body by assigning her body parts various technological functions and turning them into disposable, exchangeable, and marketable components. The latent violence involved in this kind of production and in the attendant language has perhaps been most clearly expressed in the recent serial killings of *maquiladora* workers. I will return to this parallel later.

In her article "A Cyborg Manifesto," Donna Haraway has examined the linguistic reformulation of new forms of domination in the information system in which everything is communicated in terms of rates, cost cuts, speed, proximity, and degrees of freedom. It is the language of leanness, efficiency, and competition that any corporation understands, a language that allows the universal translation of terms on which to operate worldwide.

Let me look at an advertisement for Elamex Communications—a *maquila* broker in El Paso, Texas—to see how this kind of imagery tends to reinforce certain borders of identity in reaction to the breakdown of national borders in the information society. The Elamex advertisement targets U.S. corporations that are considering transferring their labor-intensive electronic assemblies to an offshore subcontractor.[3] Predictably, the ad addresses the customer in the language he understands, that is, it speaks to the desire to cut labor costs to a fraction. It also communicates a certain quality assurance, tax shelter, speedy start-up, proximity to the HQ and U.S. markets, fast turnaround, low transportation costs, direct dial communication, and airport proximity. In these terms, labor is only a figure ($1 per hour), but it is also represented as a depersonalized, quantifiable unit and thus similar to any other incentive offered to entice manufacturers to fold their activities in a certain national space and set them up in the transational free zone.

But I would argue that the Elamex image also speaks an entirely different language. It communicates a web of psychological, social, and historical relations that are being suppressed by rational arguments for efficiency. In the border zone, everyone is being transformed into a transnational subject. Only bodies that can be marked, exchanged, turned into a commodity, and recycled will be granted the entry visa that allows a certain mobility in the transnational space. The Elamex ad operates, first of all, as a technology of surveillance in that it positions the two women it showcases within the confines of certain racial and sexual criteria. On the left, we have what we are expected to recognize as an Aztec profile with red, white, and green silk ribbons woven into the braids, and on the right we have a generic Asian profile with a pageboy hairstyle and eyeliner that simulates slanted eyes. Both women wear some sort of national folkloric outfit, so that in addition to a racialized discourse the image clearly links the women to a generic exotic/erotic national entity. This reduces them further to a geo-body, a body that is turned into an allegory for a gendered, racialized, and nationalized body of people whose national virtues are tightly linked with corporate interests. While this procedure nationalizes the female body, it also feminizes the offshore national spaces of Mexico and the Philippines—the other country to which this ad supposedly refers.

Historically, women's bodies have encapsulated the desire for conquest. In what Anne McClintock calls "the porno-tropics of the European imagination," the female figure is mapped as the boundary marker of the empire, as the mediating figure on the threshold to the feminized space of the terra incognita. McClintock explains this formation with the profound, if not pathological sense of male anxiety and boundary loss at the event of leaving the known world to explore the unfamiliar. From the outset, the feminizing of the land has been a strategy of violent containment, one belonging in the realm of both psychoanalysis and political economy (McClintock 23–24).

The Elamex advertisement foregrounds the role these historically produced desires have played in the hiring of female labor by placing the bodies of women within the fantasy narratives of colonial conquest. "Mexico beats the Far East by 10,000 miles," the headline reads. In this all too familiar scenario the two women are pitted against each other, set up to use their sexuality and femininity to compete for the favor of the male corporate employer. Their racialized, gendered figures become the articulators of the border, of the fragile line marking the fringes of the national body. According to national(ist) discourse, all disease, illegality, contamination, and poverty come from this territory. This is the most vulnerable, penetrable site, the place where anxieties of national identity tend to localize. Above all, however, U.S. customers need to be assured that the offshore female bodies are not out of control. The ad makes a point of representing a domesticated, docile, dependable, and disciplined female workforce. The worker's manicured hands meet corporate standards, her face expresses seriousness, concentration, and precision, her demeanor betrays no emotions. In short, she represents the replica, the instrument itself. Holding the semiconductor in her hand, she and it become one, her body becomes inscribed in a robotic function; the chip becomes the extension of her hand and takes the place of her upper torso. Her body has become completely technologized.

A cyber-feminist perspective tends to interpret such images of organic/mechanic border fusion as potentially empowering in that they at least shatter the attempt at creating the representation of a fixed, sovereign subject. Even though I have reservations about assigning empowering qualities to this particular Elamex image because of the gender and race clichés it cements, it is one of the rare representations of the low-end subjects of the high-tech complex. Commercial representations ordinarily feature only designers and high-end users who then benefit from the dazzling images associated with the futuristic technologies that enhance their social image and their value in the labor market. At the same time, other contributors to the industry—such as secretarial staff and maintenance personnel in the office sector or technicians and assembly workers in the manufacturing sector—are systematically excluded from these representations. More often, *maquila* women find themselves in rather dull discourses that are associated with poverty and exploitation in sociological and development contexts.

The Elamex ad operates through a double discourse by which the ostensibly opposed registers of naturalized and technologized bodies are coordinated and managed. Here, the normative link between "the female" and "the natural" is replaced not by another clear equation but rather by the disturbing identification of the feminine with an uncertain mixture of the natural and the technological. In this entanglement of mechanism and gender, the natural female body is disarticulated, inscribed onto the machine, and individually reembodied as the "hand" or the "eye" of a new corporate whole. These happen to be the body parts for which a *maquila* woman gets hired—her eyes and her fingers—because digital and microelectronic manufacturing demands both great optical precision and tactile nimbleness. But her biological components also make her fragile and vulnerable. Her eyesight is sharp enough for about eight years, but then she will have to be replaced by a fresh young worker. Her organic vision is consumed in the making of the visualization technologies our society relies on. She belongs to the process of periodic replacement by other bodies; she needs to be continuously recycled.

In his book *Bodies and Machines,* Mark Seltzer analyses with great accuracy the compulsive desire to see and make visible as one of the defining features of the body-machine complex at the end of the last century (95). This desire has driven scientists and engineers to develop an arsenal of apparatuses and instruments to multiply the potencies of the human eye. In the 1990s, Rosi Braidotti observed that the biotechnological gaze has penetrated into the very intimate structure of living matter, that it is now seeing the invisible and restructuring that which has not yet a shape (43). The desire to make everything visible is also an imperative of making things legible and governable; it expresses at once the fantasy of surveillance and the need for embodiment (Seltzer *Bodies,* 96). The optical technologies manufactured on the border convey the great importance artificial vision holds for power, now and in the future. Medical and cyber optics, surveillance, x-ray satellite technologies, telescoping, AV media and virtual technologies, identification, scanning, digitizing, controlling and simulating electronics—they all engineer the relation between seeing and power, between vision and supervision. Haraway has tried to rescue the faculty of seeing from a disembodied technological vision of the phallogocentric kind and to re-possess it for a feminist discourse. In "Situated Knowledges," she pleads for an embodied objectivity and suggests that we learn to see in compound, multiple ways, in partial perspectives that shatter the idea of passive vision in favor of the notion of the eye as an active perceptual system, continuously translating, and always accountable.

In the present situation where certain developments appear to be of global inevitability, it might be more important than ever to scrutinize what Haraway suggests are the possibilities of lived femininity/feminism in a cyborg world and to reinstall notions of singularity and subjectivity within a discourse that functionalizes the female body to the extreme. Many accounts have already related the mechanisms of containment that affect women's lives on the border. I chose not to focus on the instruments of repression in my video *Performing the Border* because I assume change is not a matter of information. But I want to describe these kinds of conditions in more detail here as they exemplify the relations between gender, body, and technology in very explicit terms.

Technologies of Control

The technologies of border and labor control installed in Juárez make the relations between vision, vigilance, power, and bodies violently obvious. Even though labor organization is legal in *maquiladoras,* every effort is undertaken to prevent it, and one of the major reasons why assembly plants prefer female workers is that they are supposed to be more docile and less likely to

Photo 4.2. Still from *Performing the Border* by Ursula Biemann, video essay, 43 min, 1999. Photograph courtesy of the artist.

organize into unions. Also, since adolescent female workers are often the only ones in the family with an income, there is much pressure from male family members on women to acquiesce to existing working conditions in order to save their jobs. The *maquila* program relies strongly on prevailing patriarchal family relations in Mexico. In recent years, the entire industrial zone has become interconnected via computer networks, and plants have established black lists containing the names of undesirable persons, starting with assassins, delinquents, and "enemies" of the *maquila,* that is, people trying to alter the conditions in the *maquiladoras.* Black lists of this sort are prohibited by law because if someone is let go by a factory, there is no chance of finding work anywhere else in the zone. Labor activist Ciprianana J. Herrera, a member of the Centro de Investigación y Solidaridad (CISO), told me that she got fired, together with two other *compañeras,* for requesting a cafeteria.[4] Their plant was located outside the industrial park, and for several hundred workers there was no place to have lunch. These workers were not even talking about forming a union, about wage policies, health hazards, or human rights. Women are afraid of losing their jobs for the slightest disobedience, of never being able to find another one again, and of imposing the consequences of their unemployment onto their families.

For the Mexican government, the *maquiladora* program is strategically important to the economy, ranking well above any other national income from oil or tourism. The government keeps a close eye on the *maquilas'* interests. And we can assume that the purpose of the strong U.S. military presence is not merely to keep "illegals" from crossing the border but also to protect the gigantic U.S. industrial investment on Mexican territory. Guillermina Villalva Valdez, a leading labor activist and academic who was extremely supportive during my first visit to Juárez in 1988, died in a plane crash on her way to Texas in 1991. In the small plane, which exploded in mid-air, presumably because of a bomb, were also four other key figures of the labor movement. Labor activities are watched closely by the networked corporate system.[5]

Time management is another efficient means of control. For practical reasons, the industrial parks are located on the outskirts of the city. Regular public transportation does not go there. At the changing of the shifts, private companies shuttle the workers back and forth between the city center and the plants at exorbitant fairs that can swallow up to a third of a woman's monthly salary. Before dawn, the *maquila* worker leaves the settlement on the periphery, walks to the bus station in the center of town, and takes a one hour bus ride out to the *maquila* to make the morning shift at six o'clock. She spends nine hours at the plant and goes back home the same way. That leaves no time to live, no time to think, no time to organize. The workers' excruciating time investment, in turn, enables the further development of technology that accelerates the lives of transnational consumers in the so-called First World.

In his essay "Going at Different Speeds," Andrew Ross identifies speed differentials and relative time scarcity as the basic principles for uneven development in the world economy. "Beyond a critical speed," he quotes Ivan Illich, "no one can save time without forcing another to lose it" ("Going" 174). In the electronically networked *maquila* system every individual is identified and profiled. Time, productivity, and the body of the female worker are strictly controlled, oftentimes by white male managers. In certain cases, this control goes as far as requiring a monthly cycle check to ensure the worker is not pregnant since pregnancy entails immediate dismissal. The reproduction of the workers' bodies is thus often strictly controlled from the moment they are determined to be productive. Thus, speedy industrialization has imposed rather violent transformations in the contradictory registers of public and private spaces, of work and plant, on the one hand, and of home and family, on the other, or more generally, of the economic and the sexual realms. In Mexico, like everywhere else, these registers have traditionally been divided along the lines of sexual difference. Women took care of the domestic sphere whereas the men in the family—father, uncles, brothers—sustained the family financially. What the borderlands has witnessed over a short period of time is the conflation of the separate spheres of the private, female, domestic space of reproduction and consumption and the public, male space of production. With the hiring of predominantly young women these traditional patterns are being forcefully transformed, but not without conflict. Not surprisingly, the female worker emerges as the central figure in this conflict as she embodies the two functions of production and reproduction. She is the embodied "problem" that needs to be contained and managed.

Since NAFTA, the border has materialized this conflict on an ever more impressive scale as the agreement assures the free flow of goods but prevents the free passing of people who produce the goods. The crossing of merchandise stands for good neighborly relations whereas the crossing of people is criminalized and policed. The border thus becomes a metaphor as well as an actual material institution that capitalizes on the differences between the economic and the sexual.

Sexualizing the Territory

One of the most striking, and perhaps most disturbing insights I gained on the border are that international labor in the South is not only feminized but also sexualized. The female workers are literally interpellated into their sexuality. Structurally speaking, a young woman in Juárez has three options: either she becomes an assembly worker, a *domestica* in a private house (if she is not sufficiently educated) or a prostitute (if she can't produce a recommendation for such a position). Yet securing a factory job is not always the end of the story. Low salaries force many women to seek supplementary income from prostitution on weekends. Sexual and labor markets interpenetrate each other within this economic order. The figure of $1 per hour I cited earlier in my discussion of the Elamex ad is also responsible for sexualizing the offshore labor market because it pushes women to become reducible to sex. The figure means that pimping takes place on a corporate level. This does not mean that transnationals are creaming off the profits from prostitution, but that they benefit from getting labor for pocket money by making women dependent upon the commodification of their bodies.

Prostitution is not just part and parcel of a tax-free consumer binge; it is a structural part of global capitalism. Since the closing down of the border and its military enforcement in the 1990s, which is described in more detail in Joni Adamson's and Claudia Sadowski-Smith's essays in this volume, competition has become more fierce between professional prostitutes and a growing number of young, often adolescent, *maquila* workers who prostitute themselves on weekends. The dynamics on the border clearly show that despite a decline in the number of customers who spend their U.S. dollars

Photo 4.3. Still from *Performing the Border* by Ursula Biemann, video essay, 43 min, 1999. Photograph courtesy of the artist.

in Juárez, prostitution is growing. In other words, prostitution is not generated by customer demand, as is usually thought, but by the women's need to produce additional income. Initially, these women offered sexual services to anyone who could pay for them, but gradually their prostitution has given rise to an entertainment industry. It is important to note that in Juárez, where prostitution emerged as a by-product of the maquila economy, the sex districts are pimp-free.

In the official discourse of the U.S. media, the border is often represented as a place of delinquency, debauchery, and prostitution, a magnet for all subjects who don't meet the moral standards of society. The media rarely wastes a word on the fact that these conditions are engineered by the *maquila* industry that implements plans designed and signed by both national administrations and by the Dow Jones people. The media, it seems, have mistaken the effect for the cause. It's not that I'm particularly interested in tracing simple causality. In an overwhelmingly complex site like the border, it can be more fruitful to record the synchronicity of events and to point out correspondences without necessarily building an overarching theoretical framework. Also, it seems inadequate to offer hasty interpretations for formations that are quite malleable and changeable. On the border, identities are constantly forming and collapsing, conforming and transgressing, so I do not want to propose easy categories for the new types of subjectivities that are currently evolving. We must not forget that there is much ambivalence within synchronicity, which is caused by conflicting interests and competing desires.

It just so happens, however, that sexuality has become a site on which desires for self-expression and control mechanisms converge violently. Thousands of assembly jobs have been created in the desert city of Juárez, and women tend to be the ones who get them. As gender relations are greatly determined by economics, the reversal of income patterns has had an immediate impact on the way women relate to men. For one thing, women have gained greater autonomy over their sexuality. On Friday at 4 P.M., when the assemblers leave the morning shift, hundreds of bars and dance saloons are already open for business in downtown Juárez. Ten years ago, on my first shoot in Juárez, the contests organized in the nightclubs were modeled along the lines of traditional gender roles. Women sucking lollipops competed on stage for male attention by performing with the most desirable body language they could muster. The winner wasn't the woman who danced the most erotically and assertively, but a girl who slowly stepped back and forth between two stage corners in aimless anticipation. It didn't seem like much, but the audience reacted with great enthusiasm to her nervous passivity that let their gaze take possession of her. She embodied all the visual pleasures Laura Mulvey spoke about in the 1970s in regard to cinema.

Today, in the dance halls the shift of buying power to young women is obvious. Entertainment mainly caters to female customers with male strip shows and male dance contests where women cheer in appreciation of men's sex appeal. Songs are dedicated to the girls from Torreón or from Durango—who make up the majority of *maquila* workers—the song lyrics often refer to female sexual desires, and the entire entertainment machine is adjusted to their pleasure. The shift in the income pattern thus also empowers women in their personal relationships. It has enabled their overt expression of sexual desires and affords the satisfaction of these desires by economic means rather than by the more traditional ones, that is, in the domestic setting through emotional or reproductive means.

Technologies of Survival

The internal diasporic movement of Mexican women from rural into the transnational space of the borderlands attests to their flexibility, resilience, and endurance. They are often still very young—13-, 14-, 15-year-olds—when they leave their families in the interior and travel long distances to work on the border. They come from towns like Zacatecas, Durango, and

Torreón on the arid central plateau and move to the Rio Grande. They are the hope of those left behind. Often they come in small groups: three or four girls of the same age and from the same town.

Upon their arrival, they won't find accommodations because municipal investments are only made for the transnationals, not for the people who work for them. So they go to the edge of the settlements, which spread far out into the Sierra, choose a vacant spot, and build a shack right into the desert sand. To do this, they use leftovers from the *maquiladoras.* Pallets will serve as walls, chemical containers become water tanks, and so forth. As Manuel Mancillas discusses in his essay on a "squatter" community in Baja California in this volume, some people call this procedure "invasion" because the migrants take a piece of land, settle down, and wait for official papers for their houses. It may be irregular, but it is the only way to obtain housing. There are vast stretches of land where mainly women live, streets of sand, no streetlights, no public transportation, and no security. It is not unusual to see young *maquila* women moving through their desert neighborhoods wearing the little flesh-tone prostheses that protect them from the excessive electromagnetic charges that run through their bodies during assembly and testing. They are electromagnetic discharge needles, and the workers wear them strapped to their wrists. Attached to them are pink curled cables that link the female body to the workbench. The *maquila* women keep the devices around their arms on weekends for fear of forgetting them on Monday morning.

It is an alien way of life: working in a corporate culture in the morning and kneading corn dough at night. The rhythm of the barren highland gives way to optimized production modes. Life on the border teaches its inhabitants to cope with contradictions, to shift out of habitual formation on a daily basis, and to operate in a pluralistic mode because flexibility is a matter of survival. It is a life in transition, and survival is a good place to start. The courage to endure the situation is a desire that exceeds power; it doesn't pretend to overcome oppression or to master it but to survive it, according to Homi K. Bhabha.[6] He proposes a philosophy of survival rather than subversion, and this seems to me to be an appropriate model at the beginning of a new century, when postindustrial systems of production and information seem to have made oppositional mass politics utterly redundant. But we should keep in mind that survival can be motivated by different living situations. While intellectuals like Bhabha and cultural producers like me may *choose* strategies of transgressions because these seem to be of particular interest to us at this point in time, others are *forced* into transgression by the oppressive situation they find themselves in, even though they may actually prefer a different kind of life.

Transgressive Identities

Even the most sophisticated technologies of surveillance have fissures and leaks, and there are holes in the border fence as well as trails that lead through the desert valley. It is here at night that women coyotes help pregnant women across the border, making use of their indigenous knowledge to avoid snakebites and dehydration. It is true that many coyotes scrupulously exploit the desperation of border crossers by stealing their money, beating and raping them, or by leaving them to die in the desert or to drown in the river. But it would be too simplified to generalize the motives of all people smugglers or even the effects of their activities. The coyotes who help pregnant women arrive safely at a U.S. hospital charge little. In the new transnational space we will be looking for their kinds of road narratives, for transgressive trajectories that express alternative desires. And even though in number and agency these nomadic and transgressive subjectivities are modest, I believe that, philosophically speaking, it is important to theorize them.

Angela Escajeda, another member of CISO (Centro de Investigacion y Solidaridad), whom I interviewed during the filming of *Performing the Border* told me the story of Concha. Angela had known Concha for about five years when she lived in a house made of leftover materials from the *maquiladora.* At some point Concha found herself abandoned by her husband and realized that there was no chance for a pregnant woman to find work in Juárez. So Concha would cross to the United States where she sold cigarettes cheaply because she didn't have to pay taxes, then she would buy merchandise in the United States, bring it back to Mexico, and put it into circulation there. She was good at avoiding U.S. immigration officers even after the militarization of the border, when attitudes towards border crossers became more aggressive. At some point in her life, without thinking about it, she began leading other people across. Her fame grew to the point where people from Central America started looking for her. She would help them cross into the United States and charge only a small amount of money compared to others. Concha often assisted pregnant women who wanted to give birth across the border and have American children so that they could obtain papers and benefit from U.S. services. Concha even ran a "service for pregnant women" that left them at the public hospital in El Paso.

Concha's narrative of transgression stands in radical contrast to the docile, knowable, manageable kinds of bodies presented in the Elamex ad. Tracing new paths that blur with the first winds, she crosses the border, moving in and out of legality. Hers is not a one-time kind of crossing with the aim of becoming someone else on the other side. Rather, she is a subject in transit, moving through the transnational zone while finding ever-new strategies to get around the prevailing power structures on her clandestine trajectory. The figure of this kind of helpful "coyote"—someone who smuggles people across the border for little money to help them make use of opportunities in the United States—expresses, in a number of ways, the sort of "new subjects" feminists and post-structuralists may be imagining. As the crosser of various cultural locations, the new subject is the mediator and constant translator of different sedimentations, registers of speech, and cultural codes. When I passed by what used to be Concha's house, she had already packed up and gone. Since she left no forwarding address, she is no longer addressable in the ordinary sense by the system of citizen control.[7] Her position is profoundly subversive through the fleeting, utterly mobile, and transitory nature of her activity and through the dis-identification with and disloyalty to any national program. With Concha's help, the pregnant, maternal body—which is ordinarily the object of great biotechnological interest and reproductive control—also has the ability to become the site of transgression. She transfers these bodies from the transnational zone, where social services are denied to them by U.S. corporate employers, to a new national space, where they can collect the benefits due to them, even though this space is ironically dominated by the same corporations.

In *Nomadic Subjects,* Rosi Braidotti reads desire as that which evades us in the very act of propelling us forward, leaving as the only indicator of who we are, the traces of where we have already been. In her terms, the nomad's identity is a map of where s/he has already ceased to be. Braidotti sees identity as a retrospective notion and nomadic cartographies as something that needs to be remapped constantly (35). Perhaps we have come to such images of subjectivity because many people, including cultural critics and cultural producers, live diasporic, transient lives even though we admit that this type of diasporic lifestyle—typical of an intellectual in the North—is not equal to the one evolving in resistance to corporate politics in the South. Yet life on the border is of a permanently transitory nature, the female diasporic subject emerges as the transgressive identity. She keeps moving back and forth between rural and urban, between rudimentary survival strategies and high technology for cyber culture, between traditional folklore and robotics. She crosses the boundaries between production and reproduction, and she circulates in these multi-layered spaces, making connections with local coalitions and international feminist networks on labor rights, environmental issues, and human rights. Braidotti's nomadism does not mean fluidity without borders but rather an acute awareness of the nonfixity of boundaries. It is the intense desire to go on trespassing, transgressing (Braidotti 35).

Serial Killings

There is another, more violent aspect to the clash between bodies, sexuality, and technology in the U.S.-Mexican border zone to which I now want to turn. Since 1995, close to 300 women have been killed in Juárez and all according to a similar pattern: Poor, slender women with long dark hair, mainly workers, rarely students, have been raped, tortured, stabbed, or strangled, and tossed into the desert. Many of them had just moved to the city so nobody knew them or claimed their bodies. Fifty women are still lying in the morgue, unidentified.[8]

Women's organizations have been formed in reaction to this acute violence in the public space.[9] Most of them interpret it as violence against women, as acts of revenge taken by a single man or several men on women who have stolen their jobs, who have started to talk back to them, who go to dance halls, and who generally challenge gender roles. The fact that the police haven't been able to solve the crimes is seen as another sign of male consent to this scenario. Feminists and human rights advocates took it upon themselves to investigate the cases and to establish a list of missing women in an effort to prove that the cases bear too many similarities to be individual crimes of passion. These groups recognize, however, that some cases constitute acts of ordinary domestic violence, which are disguised as one of the serial killings. They also understand that extreme poverty, lack of education, and economic subjugation are all conditions that prepare the ground for a criminal or several criminals to commit such horrific crimes. U.S. criminologist Robert K. Ressler, who was invited to analyze the case in Juárez, also points to drug traffic, gangs, migration, quick money, and prostitution as further conditions that might have led to these crimes—conditions no different than those in any major metropolis in the United States. But apart from the widely scattered migration settlements, Juárez is in its core still a modest border town, and serial killing is not an ordinary crime of passion. In view of the very particular constellation of economic, sexual, social, and technological factors on site, generic explanations simply don't suffice.

In the last section of *Performing the Border,* I make an attempt to establish some of these connections. I waited with this discussion until the end of my work because I believe we cannot understand the full meaning of the serial murders unless we know about the other, previously discussed components of the border. The strength of the essay as a film genre doesn't lie so much in supplying precise facts and information as in the representation of complexity and the stimulation of reflection. In the part about the serial killings, the video moves further away from a documentary format to open up more speculative spaces. Also, I should point out that at the moment of this writing, the murders have still not been solved conclusively so that whatever we might say about the number or nationality of the killer(s) today remains on the level of speculation.

In the course of my research for my video essay, I came across a recent study, *Serial Killers* by Mark Seltzer, which draws a number of intriguing connections between sexual violence and mass technologies characteristic of a machine culture. Even though he never mentions the unresolved *maquiladora* murders, the relevance of his analysis to the events in Juárez is undeniable. He traces connections between this form of repetitive and compulsive violence to the kinds of production and reproduction that make up machine culture, and he particularly relates technologies of identification, registration, and simulation to the psychological disposition of serial killers.

In his introduction "Serial Killing for Beginners," Seltzer explains that serial killers have an identity problem. "He" (with one known exception serial killers have always been male) lacks boundaries. He fails to distinguish himself from others, and this lack of self-distinction, of self-difference, is immediately translated into sexual violence since gender is the one fundamental difference he recognizes. Employing Seltzer's logic, we see that in each of the *maquiladora* murders the gendered other is made undistinguishable, exchangeable, and reduced to a pure number in a body count. Exchangeability appears to be a determining factor in the murders reported in Juárez. The victims do not only have a similar physical profile, but their bodies are also often found in different locations than their clothes, which makes their identification more difficult. Perversely, many bodies have been found wearing clothes that originally belong to other missing women. This confusion of their belongings, which might serve as markers of identification, literally emphasizes the exchangeability of the bodies. At the same time, however, the killer has left new distinguishing signs of violence on the body, often brandings and cuttings.

According to Seltzer's extensive research on serial sexual violence, a common psychological denominator of the killers lies in their undoing of identity, their desire to merge themselves with the social and physical environment. In the case of the serial killings in Juárez, we also find a strange permeability between the bodies of the murdered victims and the urban environment as well as between urbanity and nature. In Juárez, the habitat often blends into the natural surroundings, and the built reality merges with the unpaved roads. In addition, the slaying of women often happens at dawn, when the distinction between night and day is blurred and when the boundaries between private houses, unpaved streets, and the desert become even less distinguished. There are large areas where the nominal division between public and private is blurred, in part because public spaces are nothing more than private improvisation. In the early morning hours, a great number of women cross through these largely undefined spaces on their way to the *maquiladoras,* in transit between private and workspace, between desert and urban center. The assimilation of the serial killer to the milieu is nowhere better realized than in this terrain where clearly distinguishable contours are virtually absent.

In Seltzer's accounts we find a rich collection of similar analogies for the serial killer, in which persons and landscapes, bodies and technologies, the public and the private literally merge. Seltzer discusses that sexual offense and eroticized violence often cross the boundaries between the natural and the collective body, transforming private desire and pathology into public spectacle.[10] This transgression, which is characteristic of the psychological configuration of the serial killer, is performed on the site of the gendered and racialized body, and, for the murderer or murderers of Juárez, the border becomes a perfect stage for this performance. Losing the boundaries between the self and other, the serial killer(s) is/are perpetually in search of a border. For this reason, they may also be attracted to the border of their country, precisely because it signifies the boundary of a larger entity of belonging to a particular nation. Going to the border may physically express his/their mental extremity, merging his/their physical body with the national body by confusing the inside and outside.

As I have argued, the border is a metaphor for the artificial division between diverging concepts as well as a site where the blurring of distinctions takes on violent forms. On a representational level, the Elamex image I discussed exemplifies how the act of technologizing the female body simultaneously reifies identity markers of nature, gender, ethnicity, and nationality. On the material level, this process is paralleled by the robotic, repetitive process of assembly work, the intimate implication of the body in technological functions, and the association of this process with the gendered, racialized body. The serial killer(s), in turn, translate(s) the violence of these various entanglements into urban pathology, thus publicly reproducing the repetitive, disassembling, and disidentifying performance of assembly work and its discursive representations on the site of the female *maquiladora* body. Many of the victims are in fact *maquila* workers who were found wearing their *maquila* overalls and ID cards bearing the corporate logo. What the industry constructs as consumable, disposable bodies is literally tossed into the desert nearby in acts of informal "garbage disposal." In his own morbid way, the serial killer thus does nothing more than make literal and visible the prevailing discourse about *maquiladora* workers. The serial killer's identity problem makes him the perfect mediator between border discourses and institutions. He is *the* performer who makes the borderlands his stage.

There is a limit to my fascination with the murderer's pathological mind because, after all, he is killing women in raw numbers. But a discursive reflection on the killings helps to understand them as an urban pathology produced by accelerated industrialization and modernization. It also allows us to recognize how deeply the post-industrial world is implicated in the disturbing changes taking place on the border, which have a significant impact on the lives of Mexican women. To look at the border involves examining issues of representation, but their performative realization weighs above all on young Mexican women. They assemble the digital technology; their time and their bodies, down to their monthly cycles, are strictly controlled by the *maquiladora* management; and prostitution is a necessity for many in an economy that is characterized by sexual violence. Feminists in Juárez have the courage to survive and to struggle under repressive conditions, to say "no" to indifference and exploitation. I acknowledge their efforts to support other women in finding better and alternative ways of living on the border because, in doing so, they are rewriting the texts of their subjectivities and of their society as it changes and as they change it.

Notes

1. This is a revised version of a book chapter with the same title in my book *been there and back to nowhere:* 133–145.
2. Bertha Jottar, Mexican artist, introduces my video essay *Performing the Border.*
3. Elamex is the largest contract manufacturer in Mexico with annual sales of 129 million U.S. dollars (in 1998) and 17 manufacturing plants with operations in electronic and electro-mechanical assembly for the automotive, telecommunications, computer peripheral, military, and medical industries. This advertisement circulated in industrial trade magazines in the mid-eighties (see www.elamex.com).
4. Cipriana Herrera works for CISO, the *Centro de Investigación y Solidaridad* (Center for Research and Solidarity). CISO is a counseling and advocacy non-governmental organization without an office; the members drive to various neighborhoods and advise people on labor laws.
5. Guillermina Villalva Valdez was founder of COMO, the *Centro de Organización para Mujeres Obreras* (Center for the Organization of Women Workers), where women workers were educated and politicized.
6. I am quoting here from Homi K. Bhabha's lecture in September 1999 at the School of Art and Design, Zurich, in the context of an exhibition and symposium on cultural practice in South Africa.
7. Concha's border crossings have also been the subject of *Power of Place,* a CPB program of the World Regional Geography series, produced in 1994. According to the producer's recent update on her story, Concha has, in fact, gone to work for a *maquiladora.* She was apprehended by the U.S. Border Patrol one too many times. Threatened with serious jail time, she thought about what it would mean for supporting her kids and opted to give it up.
8. Conversations with Judith Galaza of CICH, a human rights organization.
9. Three groups that were recently formed in reaction to the *maquiladora* murders are a circle of nine women journalists who regularly come together to discuss their research on the killings: "8th of March," a small group that engages in presswork and organizes the mothers of missing girls; and CICH, a human rights organization.

10. To back up these statements, Mark Seltzer quotes Anne Rule's "Serial Murders: Hearing on Patterns of Murders Committed by One Person, in Large Numbers with No Apparent Rhyme, Reason, or Motivation," Testimony before the U.S. Senate, July 12, 1983 (Washington, D.C.: U.S. Government Printing Office, 1984).

Works Cited

Bhabha, Homi K. Presentation at the School of Art and Design, Zurich, September 1999.

Biemann, Ursula. *been there and back to nowhere: gender in transnational spaces.* Berlin: b_books, 2000.

——. *Performing the Border.* (43 min.): 1999.

Braidotti, Rosi. *Nomadic Subjects, Embodiment and Sexual Difference in Contemporary Feminist Theory.* New York: Columbia University Press, 1994.

Haraway, Donna J. "A Cyborg Manifesto" and "Situated Knowledges," *Simians, Cyborgs, and Women: the Reinvention of Nature.* New York: Routledge, 1991: 149–181 and 183–201.

McClintock, Anne. *Imperial Leather: Race, Gender and Sexuality in the Colonial Context.* New York: Routledge, 1995: 23–24.

Mulvey, Laura. "Visual Pleasures and Narrative Cinema," *Art After Modernism, Rethinking Representation.* Ed. Brian Wallis. The New Museum of Contemporary Art, New York, 1984.

Power of Place. CPB Program. 1994.

Ross, Andrew Ross. "Going at Different Speeds," *Readme! Filtered by Nettime.* New York: Audonomedia, 1999.

Seltzer, Mark. *Bodies and Machines.* New York: Routledge 1992.

———, *Serial Killers: Death and Life in America's Wound Culture.* New York: Routledge, 1998.

2. Border Communities

Fan Letters to the Cultural Industries

Border Literature about Mass Media[1]

Claire F. Fox

Llegaba el Domingo
Llegaba el domingo y llegaba el pánico.
Veinticinco horas y nada que hacer.
Tendríamos que decidir entre los tres cines.

En el De Anza enseñaban películas
francesas y no podríamos entrar porque
estaría henchido hasta los pasillos
con casados.

En la Ensenada estarían exhibiendo
una variedad de películas—
LA VIDA DE PANCHO VILLA
LA MUERTE DE PANCHO VILLA
EL RETORNO DE PANCHO VILLA
LA REVANCHA DE PANCHO VILLA
EL CABALLO DE PANCHO VILLA
LA PISTOLA DE PANCHO VILLA
LA CABEZA DESAPARECIDA DE PANCHO VILLA
LA CABEZA ENCONTRADA DE PANCHO VILLA

Sólo quedaba el Cine México donde
pasaríamos la tarde del domingo
maldiciendo a John Wayne por cada
yaqui, apache y mexicano que mataba.

Sunday Arrived
Sunday arrived and panic arrived.
Twenty-five hours and nothing to do.
We would have to decide between the three theaters.

At the De Anza they showed French films
and we wouldn't be able to get in
because it would be crammed up to
the aisles with married men.

At the Ensenada they would be showing
a variety of movies—
THE LIFE OF PANCHO VILLA
THE DEATH OF PANCHO VILLA
THE RETURN OF PANCHO VILLA
THE REVENGE OF PANCHO VILLA
THE HORSE OF PANCHO VILLA
THE PISTOL OF PANCHO VILLA
THE DISAPPEARED HEAD OF PANCHO VILLA
THE FOUND HEAD OF PANCHO VILLA

There was only the Cine México left
where we would spend Sunday afternoon
cursing John Wayne for every Yaqui,
Apache and Mexican that he killed.

Introduction

"What movie should we see this Sunday afternoon?" This is the weekly dilemma faced by the *fronteriza* narrator of Gina Valdés's poem, "Llegaba el Domingo" (1986).[2] None of the choices is particularly appealing: The Ensenada theater shows stale Mexican movies about Pancho Villa, and the De Anza, which features French movies, is overrun by married men. That leaves the Cine México, where the narrator finally ends up spending her Sunday afternoon, "cursing John Wayne for every Yaqui, Apache and Mexican that he killed." The theme of the spectator's conflicted relationship to visual mass media appears repeatedly in the writings of authors who live and work in the border region. The border served as a source of narrative raw material for the Hollywood and Mexico City-based film industries, especially during the first half of the twentieth century. At the time, the U.S. western and *film noir,* and Mexican *cabaretera, charro,* and revolutionary pictures, used the border to refer to a world where the reigning values and social systems were contrasted to the interior of each country.[3] But the Mexican cinema's Golden Age, with its strong genre and star system, declined in the early 1950s, after which the industry experienced periods of resurgence that were usually characterized by the production of "quality" pictures for urban, cosmopolitan, and international audiences. Valdés seems to capture both the rise and fall of the Mexican industry in her poem's parodic reference to Pancho Villa movies. The revolutionary melodrama had been a stock genre of the Golden Age of Mexican cinema, and *norteño* leader Pancho Villa was of course a major icon of that genre, but the low-budget cycles with *fronterizo* themes, such as *narcotráfico* are more characteristic of the early 1980s. Meanwhile, the Hollywood industry continued to grow and expand in foreign markets through the second half of the twentieth century in a new form of Manifest Destiny that is suggested through Valdés's invocation of John Wayne as a representative of the U.S. cinema in Mexico. Spectators in the border region at mid-century thus found themselves located at the crux of two national media industries, the one waxing and the other waning, but neither of which seemed to address the subtlety and diversity of border social structure, ethnic composition, language use, and everyday life.

In this essay I argue that there exists a binational film culture unique to the U.S.-Mexico border region and born out of the region's uneven configuration of mass media outlined above. I propose to discuss "Llegaba el Domingo" by Valdés (Los Angeles/Ensenada/San Diego, 1943–), and selected prose works by Gabriel Trujillo Muñoz (Mexicali, 1958–), Norma Elia Cantú (Nuevo Laredo/Laredo, 1947–), and Dagoberto Gilb (Los Angeles/El Paso, 1950–) as examples of how literary intellectuals have engaged critically with that binational film culture. Valdés's poem, for example, mediates one narrative at the same time that it produces another. In my view, this shift from film to literature, and from consumption to production, is a source of tension and ambivalence that calls for further exploration of *fronterizo/a* intellectuals' position toward the cultural industries. The authors whom I will discuss all grew up in the 1950s and 1960s in environments saturated by U.S. and Mexican mass media, and they witnessed the shifting balance of power between Hollywood and Mexico City. Their work intercalates binational film references with references to other watershed marks of their generation, such as the 1968 Mexican student movement, rock music and hippie counterculture, the Viet Nam War, and the Chicana/o movement. In general, their treatment of the cinema is complex, marked at once by a desire to emplot the *fronterizo/a* in the movies—as director, scriptwriter, or star—while at the same time exposing the processes of film production to critique the manner in which the border and its inhabitants are implicated in mainstream narratives. One consequence of this perspective is a recurring thematic emphasis on the border's untold stories and on other ways of telling them.

With its oppositions between spectator and screen, heroes and villains, and reality and fiction, the real and virtual spaces of film culture offer an allegorical framework to reflect on territorial divisions in the context of U.S.-Mexico relations, and to problematize the stability of those divisions from a local perspective. Perhaps the sense of panic felt by the narrator of "Llegaba el Domingo" stems from the fact that her leisure time is a return to rather than an escape from imperialist domination of Mexico by the United States, and political and market-driven attempts to naturalize the borderline. Faced with a menu of masculinist, national products, she and her companions end up improvising a provisional, antagonistic relationship to the hero of the U.S. picture that stresses their own self-identification according to racial and ethnic markers in addition to national identity. They choose to occupy U.S. territory, as it were, but only contentiously and vociferously, as advocates for Wayne's victims.

Valdés's poem charts the intricate pre-history of the act of consumption, those "tactics" of individuals in the face of capitalism's limited options that are described by Michel de Certeau in *The Practice of Everyday Life.*[4] I do not wish to rest with an interpretation of this poem as an example of resistant spectatorship, however, for to do so would be to overlook its analytical and performative qualities. As studies of the reception of popular culture increasingly turn toward ethnographic models based on empirical research, active spectatorship has emerged as the norm, rather than the rarity that was once praised over its feminine and untutored other, passive spectatorship.[5] Media critics such as Judith Mayne and Ien Ang have further cautioned against the voluntarism of committed scholars who celebrate the fact of heterogeneous reception of media as evidence of resistance, and resistance in turn as intrinsically progressive.[6] Between the essentialist pitfalls of radical empiricism and de Certeau's declaration that under late capitalism, "marginality is becoming universal" (xvii), I maintain that it is worthwhile to trace the contours of collective regional discourses of reception; these in turn serve as points of comparison in charting transnational flows of media and capital.[7] The texts that I examine here do not offer case histories of reception itself, but rather are *representations of reception,* in which a utopian vision of binational film culture emerges to challenge the Hollywood and Mexico City cultural industries. My focus on cultural production in this essay is a response to contemporary trends in reception studies, which I find too often celebrate consumption as a form of resistance, while neglecting to examine the mass media's conditions of production and structures of exclusion.[8] In my analysis of border literature about cinema, I try to steer a middle path between the top-down productionist orientation of Frankfurt school criticism and the neopopulist tendencies of contemporary reception studies.

In this light, one striking dimension of Valdés's poem is its playful overview and critique of existing film production. "Llegaba el Domingo" cites real movie theaters in Ensenada, Baja California Norte, where the author spent her early childhood (Sánchez 277), and her tripartite description of local film fare echoes the descriptions of impoverished national film cultures that intellectuals associated with the New Latin American Cinema movements of the 1960s and 1970s observed throughout Latin America. On the one hand, they found a low-quality national product, and on the other, a high-budget spectacle from the United States; the alternative was an aestheticist European cinema. By infusing the European *atelier* model of production with Latin American political content, they offered their own innovative movies in support of national liberation struggles in Latin America and the third world. In a similar spirit, though with different values and spatial imaginary, I would argue that Valdés's poem, through its manifest dissatisfaction with that which is available, implicitly calls for a locally produced, gender inclusive, binational cultural production, a type of production that is satisfied in oblique fashion through writing and publishing in a bilingual format. In contrast to her Latin American peers, it is interesting to note that Valdés eliminates Europe as a middle term, in order to focus on the more immediate confrontation between the United States and Mexico in the border region.

On another level, "Llegaba el Domingo" portrays antagonistic identification with Hollywood narratives and heroes as a barometer of political consciousness, and in this respect it reiterates a trope found in many works of neo-colonial literature. Existential rites of heterodox identification at the movies appear in the writings of several generations of anti-imperialist intellectuals who were raised on a diet of classical Hollywood productions. Carlos Fuentes recalls a similar incident to that of the Valdés poem, when as a child growing up in Washington, D.C. during the period of the Cárdenas oil expropriations, he found himself shouting "Viva Mexico! Death to the gringos!" at a screening of *Man of Conquest* (Fuentes "How," 8). Similarly, Ella Shohat and Robert Stam chronicle the galvanizing effect that youthful viewing of *Tarzan* movies had on intellectuals as diverse as Haile Gerima, Edward Said, and Frantz Fanon (157). In border literature by Chicana/o authors, the rite finds its counterpart in anti-racist struggles within the U.S. John Retchy's autobiographical essay, "El Paso del Norte," for example, employs overcoded language to describe a moment of self-recognition as Chicano while passing for white at a segregated "hick cinema" in Balmorhea, Texas: "the man taking tickets said, 'You boys be sure and sit on the right side, the left is for spiks.' So I said I was on the wrong side and walked out."[9]

It is interesting to note that Anglo-American feminist film theory, dating from the 1970s and elaborated during the "early confrontational moments of [the feminist] movement," also posited spectatorial identification in either/or terms (Mulvey, cited in Mayne 30). Laura Mulvey's pioneering essay, "Visual Pleasure and Narrative Cinema" (1975), for example, described the pulse of classical Hollywood narrative as a dyadic oscillation between the active male gaze and the passive female object, and thus offered female spectators limited identificatory options.[10] These particular examples of dualistic identifications from anti-colonial, anti-racist, and feminist movements suggest that scenarios of spectatorial identification are informed by social struggles in which partisanship is deemed crucial. Rather than provide a window on cognitive processes or a reflection of real audience behavior, I would argue that "Llegaba el Domingo" promotes a vision of the movie theater as a semi-public institution where collective identities may be articulated through participatory spectatorship.[11] The young women's jeers at John Wayne are neither futile nor are they revolutionary, but they rather affirm a regional, collective identity formation that might also be mobilized in other contexts.

Recent scholarship about spectatorship has tended to shift attention away from the dynamic between screen and spectator toward extra- or para-cinematic engagements with "film culture" in a broadly-defined sense (Mayne 29). Movie star discourse and fan culture are two hot spots of contemporary reception studies that provide links between the viewing experience and other manifestations of film culture. As Miriam Hansen has observed, the appearance of a star in any given movie invites spectators to think beyond the narrative at hand:

> Because the star is defined by his or her existence outside of individual films, by the publicity that surrounds his or her professional and "private" personality, the

> star's presence in a particular film blurs the boundary between diegesis and discourse, between an address relying on the identification with fictional characters and an activation of the viewer's familiarity with the star on the basis of production and publicity intertexts. (246)

Appropriating themselves of local historical anecdotes about movie stars and film production in the border region, Gabriel Trujillo Muñoz, Norma Elia Cantú, and Dagoberto Gilb explore the tensions indicated by Hansen creatively and productively. In their work diffuse, everyday encounters with film culture outside of the theater offset the urgent, politically informed scenario of identification found in Valdés's poem.

Production "in" and Production "on" the Border

Before I discuss the works that will be the focus of this essay, it is important to survey briefly the type of film and video production that does exist in the border region. The border has never been the site of a centralized production infrastructure dedicated to making feature-length movies destined for broad audiences. Instead there have been sporadic and short-lived attempts to establish production facilities in cities such as Brownsville, Tecate, and Tijuana. There have been occasional independent feature films that capture regional markets, such as Efraín Gutiérrez's *Please Don't Bury Me Alive!* (1976), a South Texas production that "single-handedly broke Mexico's monopoly over the 400 Spanish language theaters in the United States," and inspired the Chicano film movement (Noriega). And finally, there is a variety of small-scale film and video productions, targeted at specific local audiences from activists to the avant-garde art scene.[12] Perhaps the most commercially successful form of visual media production in the region is the predominantly Mexican, low-budget industry specializing in themes such as *narcotráfico* and cross-border migration. Having risen to prominence in the 1970s, this genre is now marketed primarily on video to migrants from the Mexican interior residing in Mexico and the United States (Iglesias Prieto, I).[13]

Alongside this range of local production, the national film industries exert a strong presence. Tijuana and its environs were promoted in Alta California as a "Hollywood south of the border" from the 1920s through the 1950s. Prohibition initially drew U.S. pleasure seekers to the border states, and later Tijuana's glamorous night life and tourism industry attracted visits from both Mexican and U.S. movie stars. They left behind *crónicas* and urban legends that still circulate among the area's older residents, as well as faded photographs that adorn the walls of local restaurants and night clubs (Girven; Proffitt 185–216; Trujillo Muñoz, *Imágenes* 6–46). A more substantial manifestation of the cinema on the border stems from the impact that movies filmed in the region have had on the environment and local populations. The 1997 blockbuster *Titanic* (dir. James Cameron), shot on location in the fishing village of Popotla, Baja California Norte (south of Tijuana), is only the most recent example. While local authorities praised the jobs that *Titanic* brought to Baja California (Espinosa), other Baja Californians viewed *Titanic* as another *maquiladora,* bringing with it low wages, environmental degradation, and the dislocation of local populations.[14] Gabriel Trujillo Muñoz conveys this perspective in his history of the cinema in Baja California. "In Mexicali as well as Tijuana and Ensenada," he writes, "flyers were distributed announcing, 'Welcome aboard! *Titanic* seeks people of all ages who have European features. White complexion. Color of hair or eyes not important. If you would like to participate in the film as an extra, come to casting.'" (*Imágenes* 95). As the film production was underway, two site-specific art projects featured in *inSITE 97,* the well-known exhibition held concurrently in Tijuana and San Diego, invoked *Titanic* critically, contrasting the space occupied by the enormous movie set to the living and working areas of surrounding residents.[15]

The Valdés poem and local responses to *Titanic* demonstrate that an intertextual and interarts approach is essential to charting film culture in this region that is saturated with cinema but nevertheless lacks its own large-scale film industry.[16] Above other forms of cultural expression, I would argue that literature emerges as the site of a peculiar, contestatory dialogue with the national film industries that illuminates local conditions of film reception and problematizes existing models of studying national cinema, as well as demonstrates how literary forms themselves have developed in response to the mass media.[17] As Yuri Tsivian found when he scanned Symbolist poetry for his study about the reception of early cinema in Russia, "We start by contrasting cinema to traditional narrative; then we come to discover cinema within literary discourse. One ought to think of [film and literature] as an ensemble rather than as consisting of discrete art forms" (12).

"Border Hotel": An Archaeology of Cinema on the Border

Gabriel Trujillo Muñoz's short story "Border Hotel" (1993) is a good starting point for this discussion because it rehearses many of the stereotypes about the border that have prevailed throughout the twentieth century. The story catalogs a repertoire of roles that the region has fulfilled for the United States and its citizens, from being a site of sexual tourism, to a magnet for hippies and existential heroes, to a bastion of U.S. militarization. The narrative consists of four vignettes that transpire within the same hotel during different historical moments. Each vignette turns around a dialogue between two characters from south and north of the border. In the first scene, circa 1922, a self-absorbed and self-loathing Rudolph Valentino spends the night with a fifteen-year-old Mexican virgin, whose sexual services he has contracted through the hotel manager. The next vignette takes place in the hotel bar circa 1943, where Raymond Chandler's hard-boiled private detective Philip Marlowe shares a drink and confidences with the leftist Mexican journalist José Revueltas. In the third vignette set around 1967, Jim Morrison, lead singer of The Doors, lies in his "mushroom-shaped room" (165), tripping on peyote that he has obtained from Don Juan, the Yaqui shaman popularized by author Carlos Castañeda. In the final vignette, circa 1990, former Panamanian President Manuel Noriega sits poolside with an unnamed U.S. general discussing the recent U.S. military invasion that removed Noriega from office. The final lines of the story place all of these characters in the same space and time. As Valentino and the girl look out their window at Noriega and the general boarding a helicopter, Revueltas and Marlowe spill out into the street singing drunken *corridos,* and Jim Morrison lies in his room hallucinating.

Trujillo Muñoz is a native of Mexicali and author of a diverse and prolific body of poetry, plays, novels, short stories, criticism, and videoscripts. He is also one of Baja California's foremost *cronistas,* having compiled ambitious volumes about the region's arts, culture, and history. Trujillo's 1997 history of the cinema in Baja California, *Imágenes de plata* (*Silver Images*), illuminates his selection of characters for "Border Hotel." Here one learns, for example, that Rudolph Valentino married his second wife, Natasha Rambova, in Mexicali in 1922, while filming *The Sheik* in nearby Yuma, Arizona. The newlyweds spent their honeymoon at Mexicali's Hotel Internacional (7–8).[18] Jim Morrison also had a cinematic connection in Baja California. He wrote a screenplay for a remake of Ida Lupino's 1953 border thriller, *The Hitchhiker,* but only a few scenes were ever filmed (40–41). And, José Revueltas, Dashiell Hammett, and Raymond Chandler all authored stories with cross-border themes.[19] Although these anecdotes do not appear in "Border Hotel," the story's close ties to *Imágenes de plata* suggest that it is an attempt to revisit the historical data creatively, in a manner that disrupts linear chronology while still privileging place. In other words, "Border Hotel" reclaims the historical anecdotes and presents them as part of a pervasive and continuous pattern of inter-American encounters in the border region from the point of view of the Mexican host culture.

"Border Hotel" portrays its fictional, historical, and anonymous characters with equally schematic brush strokes. Quotidian aspects of the celebrities' private lives are stressed over their glamorous public images, and each of the four encounters turns around the revelation of a secret that destabilizes the characters' composure, albeit momentarily—Marlowe confesses to Revueltas that he has just killed a man, for example. On the one hand, the emphasis on the off-screen lives of media figures is an anti-cinematic gesture, which undercuts their screen image in favor of the type of local history outlined above. But at the same time, the story's heterodox portraits only enrich the layers of gossip and legend that surround figures such as Valentino and Morrison,[20] and thus they reinforce traditional notions of the border as a place where people may slip out of character and assume new identities through encounters with the "other."

The Valentino vignette is the part of the story where the themes of intercultural encounter and identity formation are most fully developed. While gazing out the window at other gringo tourists accompanied by prostitutes, Valentino reflects on his own exotic image and his ethnic difference with respect to other leading males (hence, his proximity to Hollywood stereotypes of Mexicans). To no one in particular he laments, "But I don't matter: What matter are the images that I am, the characters I assume with happy savoir faire and defiant daring: the buccaneer, the prince, the Arabian sheik, the hero who makes love with all the women, the one who always emerges triumphant from all the intrigues" (160). And then he reveals to his companion that he needs her because, "I can't live without having a public that sees how I act . . . It depresses me to get up in the mornings and discover that I'm nothing more than a dirty, greasy Italian not worth a shit, one more man who can be ignored like a minor character in whatever cheap movie" (160). As Valentino rambles, the girl silently makes fun of him and recalls her own acting lessons: "My mother told me . . . that I have to pretend that I like everything he does with me and it excites me" (160). This first vignette is the only one featuring a sexual encounter and a female character. It invokes the well-worn trope of characterizing U.S. imperialism in Mexico through the gendered lens of *malinchismo.*[21] At the same time, it troubles the facile and essentialist nature of this explanation by revealing the self-consciousness of each character's performance. This encounter becomes a dialogue only for the reader, who is in a position to apprehend the plenitude of the prostitute's silence in response to Valentino's soliloquy.

The Morrison/Don Juan and General Noriega vignettes repeat the unequal power relation that marks the Valentino/Prostitute encounter. If there is the possibility of the border fostering a peer relationship between North and South, this occurs in the homosocial bond that develops between the story's two intellectuals, Marlowe and Revueltas, both of whom share a private code of justice in the face of overwhelming institutional corruption. This vignette reveals Trujillo's affection for the hard-boiled detective,[22] and it also revindicates the figure of the critical (literary) intellectual who searches for the "truth which others hide" (161). Revueltas, the epitome of the Mexican oppositional literary figure, served as an inspiration to the writers who emerged from the 1968 student movement and those of subsequent generations. From a border perspective, the union of writer and detective is a useful one, for it rather seamlessly valorizes figures from both Hollywood movies and Mexican literature. On a greater scale, "Border Hotel" repeats this gesture by championing the literary at the level of plot and favoring the cinema in its formal aspects. The story's format resembles that of a shooting script: Its descriptive passages are limited to brief statements that give the time, location, and movements of characters; the events transpire in the present tense; and, dialogue is presented without speech markers (in a style similar to the novels of Manuel Puig). At the conclusion of the third vignette, there is even a moment in which the reader is addressed as a spectator by the story's trickster figure, Don Juan, who "turns toward us and winks" (163). This invitation of complicity between spectator and character evokes early U.S. cinema, with its frontal organization of action, emphasis on spectacle, and inclusion of diegetic narrators as mediators of the viewer's look. That mode of direct address would eventually be suppressed in favor of the "classical Hollywood" model that was constructed around the viewer's unacknowledged voyeurism, and where an actor's direct look into the camera would be read as breaking the illusion.[23] In "Border Hotel," Don Juan's wink at the audience asks the reader to consider the orchestration of the story's encounters. The vignettes' ironic and archetypal nature suggests a model of viewing movies as "shows" rather than as engrossing narratives, and in so doing, the story sustains an attitude of both fascination and disbelief toward the cinematic image.[24]

Canícula and Fan Culture

Whereas "Border Hotel" focuses on the private lives of stars, Norma Elia Cantú's "fictional autobioethnography" (xi), *Canícula: Snapshots of a Girlhood en la Frontera* (1995), weaves fan culture into the fabric of everyday life among the Mexican and Mexican-American communities of Laredo, Nuevo Laredo, and surrounding cities of the Lower Rio Grande Valley. *Canícula*'s narrative unfolds through over eighty stories and twenty black-and-white photographs that trace significant moments in the life of its protagonist, Azucena Cantú, dating from her birth in the late 1940s to her adolescence in the 1960s, couched within a framing narrative set in the 1980s. The stories recounted in *Canícula* are not presented in chronological order nor are they all illustrated, but they are all motivated by meditation on photographic images, as though "haphazardly pulled from a box of photos where time is blurred" (xii). In some cases the snapshots that find their way into the text have been visibly altered or do not jibe with the visual details of the stories. This disjunction between the verbal and visual registers works in concert with other memory-related oscillations in the narrative, such as the juxtaposition of the protagonist's mature and juvenile voices, word play between English and Spanish, and abrupt changes of time and location from one episode to the next. And yet in spite of these explicit manipulations of linear narrative, an ethnographic rhythm provides a connecting thread among the stories. The "autobioethnographic" impulse of *Canícula* is concerned with highlighting those practices and social institutions that forge Azucena's identity and development, ranging from oral histories and kinship relations, to language use, religion, gender relations, education, holidays, literature and popular culture, medical practices, food, clothing, and political struggle.

For Azucena and her family and friends, the mass media are not perceived as invading influences from North or South, but rather they integrate themselves easily within existing social structures. In fact, it is Azucena's engagement with fan culture as a young girl that produces and nurtures the development of her writerly consciousness. Binational movie culture is also one of the media that seems to foster bonds among women and facilitate communication among generations. As Azucena's paternal grandmother tells her the story of the family's multiple "repatriations" fleeing war and the draft, for example, Azucena references an image from the U.S. cinema to visualize her grandmother's oral history: "Mamagrande tells me stories of crossing the river 'en wayin'—and I imagine a covered wagon like in the movies" (17). Her first description of her relationship with her mother revolves around their shared pleasure in the Mexican women's magazine, *Confidencias,* which they buy one afternoon in Nuevo Laredo while the men in the family get haircuts and shoe shines: "I'll read [*Confidencias*] a escondidas, during siesta time. Hiding in the backyard, under the pirul, I'll read 'Cartas que se extraviaron,' and pretend the love letters are for me, or that I wrote them, making the tragic stories mine. I pretend I'm a leading star—María Félix, Miroslava, Silvia Pinal. During recess, I retell the stories to Sanjuana and Anamaría, embellishing them to fit my plots" (9). This passage introduces two role models that will play a central part in Azucena's development: The movie star, here identified with divas of the Mexican "Golden Age" cinema, and the writer, identified with the feminine romance genre. In subsequent stories, Azucena continues to customize the narratives of mass media, inscribing herself and other family members in the language of cinema, "to fit the plots" of her own life. She describes her mother several times as "movie-star beautiful" with "María Félix eyes" (107; see also 42, 48), and extends similar compliments to her proud Tía Piedad and her renegade cousin Elisa (85, 78). Through personalizing the language of movie culture and making public that which had remained the domain of home and family, *Canícula* dissimulates a local movie production, where snapshots become publicity stills, family members are stars, and the events of a young woman's daily life on the border are worthy of representation to a mass audience.

The force of the English language, required by the U.S. public school system and a life increasingly centered in Laredo, eventually superimposes itself on Azucena's early Spanish-, familial-, and feminine-identified engagement with mass culture. The shift occurs somewhere around the age of twelve, when she is about to enter junior high school and a repertoire of images from U.S. television and movies plants itself firmly in her consciousness, reconfiguring earlier memories, as in the following description of a photograph: "I'm wearing my cousin Tina's hand-me-down dress . . . The wide belt with the cloth covered buckle reminds me of a dress Audrey Hepburn wears in *Roman Holiday* or some such movie I saw many years later, for at that time we only saw Mexican movies at the Cine Azteca or Cine México in Laredo or at the theaters in Nuevo Laredo" (26). Around this point, Azucena's Mexican cousins begin to react differently to her during her summer visits to Monterey. In their eyes, she has become a "pocha," de-Mexicanized, and that makes her all the more "homesick for my U.S. world full of TV—Ed Sullivan and Lucy and Dinah Shore and Lawrence Welk" (22). But Azucena's growing biculturalism also gives her an increasingly important role in her own household, where she acts as translator of the news and entertainment programs for the elders and presides over the selection of U.S. and Mexican programs for the neighborhood children who congregate in the living room to watch cartoons and cowboy shows each evening (120).

Azucena's involvement with television reinforces her earlier identification as narrator and mediator of narratives for others. As in the previous example of *Confidencias,* it is through fan culture that she returns to her fantasy of being a writer. This time the desire is awakened by a locally produced cowboy show starring an Anglo Cowboy Sam and the blonde station manager's wife as his cowgirl sidekick. An ardent fan, Azucena enters a story-writing contest sponsored by the show and receives only an honorable mention for this first foray into creative writing.[25] Her recollection of this experience intercalates mature insights about the story-writing contest with grade school memories of square dancing while dressed in a cowgirl costume:

> I received the story back with the judges' comments, which I have erased from my memory, but one thing I remember about the story is that it had no female characters and the cowboy, the hero, saved the day for his friend and killed the bad guys in a shoot-out—not very creative and quite predictable given the models in the form of movies and shows I was watching. And all the while my uncles in Anáhuac herding cattle and being real cowboys, my aunts living out stories no fifties scriptwriter for Mexican movies or U.S. TV ever divined. Peewee, Angelita, Sanjuana, and I, our partners stand behind us, second graders square dancing, counting—one, two, three, four, under, one two three four, under—as the music blares over the loud speaker.

The overarching theme of "Cowgirl," the story in which this passage appears, is that of indoctrination to "American" culture and values through television and rote learning in primary school. The figure of the cowboy appears prominently in both of these arenas. Earlier in this story, Azucena contrasts the dashing, romantic Mexican *charros cantores* (singing cowboys) that she recalls from her early encounters with Mexican movies, to the Indian-fighting U.S. cowboy heroes that she finds in television series such as "The Cisco Kid" and "The Lone Ranger," and whose exploits she and her friends faithfully reenact every day after school. In hindsight, she recognizes that neither of the Mexican nor the U.S. cowboys correspond to her own first-hand knowledge of border cowboys. Her fascination with the U.S. cowboy genre is a particularly uneasy one, however, for its formulae provide heroic roles only for Anglo men and belie a history of Anglo violence in South Texas directed at Indians and Mexicans. This sinister association is underscored once again toward the end of *Canícula,* when Azucena recalls the rape of her childhood friend by a local Anglo merchant. Sent by her mother to purchase something from his store, Azucena cannot make eye contact with him and sees only his cowboy boots (120). To that day, she reveals, she cannot bear the sight of cowboy boots.

Like the narrator of "Llegaba el Domingo," Azucena's movement toward political engagement, intellectualism, and independence becomes clear through her rejection of certain mass media icons, such as the cowboy. Azucena likewise brings the romance genre full circle from its first appearance in *Confidencias* magazine. *Canícula*'s penultimate story describes Azucena's last summer spent in Monterey before graduating from high school. There she falls in love with a young medical student named René who offers her the opportunity of living out one of the movie fantasies that inspired her as a girl: "And I have a 'pretendiente' whose beautiful words and light green eyes make me dream I'm the character in *Espaldas Mojadas,* the pocha who marries the Mexican wetback leaving her border and U.S. existence to become a Mexican with David Silva" (127). The 1953 Mexican movie to which Azucena refers was produced in response to the Bracero Program (1942–1964) and was surrounded by a liberal nationalist arguments in favor of "repatriating" Chicanos/as to their "native" country (Fox 97–118). All too soon, René extends Azucena's narrative beyond the happy ending of *Espaldas Mojadas* (*Wetback*), which concludes with the lovers crossing the border into Mexico. René foresees a wedding, a honeymoon in Europe, a house, and many children—and that is when Azucena refuses to go along with his plot. Through Azucena's narratives not chosen—the Anglo Western and the Mexican romance—one discerns another narrative that incorporates elements from both traditions and in turn refracts them through the lens of a local social structure invisible to the producers of national mass media.

One other ramification of Azucena's investment in fan culture is her finely tuned sensibility to the stars' looks, that is, the hairstyles and fashions associated with them, and her desire to endow herself and other women with those looks. In her adolescence, she even becomes a hairstylist, a specialist in creating looks for others. At the same time, however, Azucena repeatedly reflects on her own unassimilable look, that is, the gaze that she projects outward through the many photographs of herself included in the text.[26] Azucena's appearance as she passes from infancy to adolescence is mercurial. Her descriptions of her own physical and emotional development are complex for their simultaneous insistence on her own sameness and difference with respect to previous moments in her life: "In the photo stapled to my U.S. immigration papers, I am a one-year-old baldy, but the eyes are the same that stare back at me at thirteen when I look in the mirror and ask, 'Who am I?' and then go and cut my hair standing there in front of the mirror, just like Mia Farrow's in *Peyton Place*" (21). This simple and poignant question—"Who am I?"—echoes throughout *Canícula*'s exploration of the multiple thresholds between girlhood and womanhood, English and Spanish, and the United States and Mexico. The temptation to read the reference to U.S. immigration papers and Mia Farrow above as evidence of a growing identification with the United States or with a particular type of mass mediated femininity, for example, is countered by the illustrations that accompany the story in which it appears—two obviously doctored Mexican citizenship documents complete with mug shots issued to "Azucena Cantú" at age one and age sixteen, respectively. In the end, *Canícula* does not resolve the dual attraction for writer and star expressed by Azucena at the outset of the text, and maintained through the narrative's visual and literary registers. Azucena's narrative spans both the verbal and visual descriptions, however contradictory they may seem, and perhaps more importantly, it is fabricated precisely in those moments where the processes of cross reference, representation, and identification are shown to be faulty or imperfect.

"The Magic of Blood" and the Heart of the Industry

Dagoberto Gilb's "The Magic of Blood" (1993), a short story from his collection of the same title, also builds tension between word and image, but in contrast to *Canícula,* it marshals words to cast doubt on Hollywood's glamorous image. Through ironic, understated prose, the story describes a pilgrimage of sorts undertaken by the narrator and other members of his family from an unspecified city in the Southwest to Los Angeles. The story ultimately suggests that greater proximity to the cultural industries does not make much of a difference for those whose access to mass media production is already highly restricted. In this case, the narrator's journey to the heart of the U.S. film industry is also a quest for knowledge about his own family, specifically about his great-grandmother who left a small town in Mexico for Hollywood when she was sixteen years old, and who, it is rumored, had a career as a film actress. The narrator declares, "What mattered, to all of us, was this one glamorous, and verifiable, detail: she had a Hollywood address. It was like believing that there was magic in our blood" (66). He finally gets his chance to find out more when he and his sister are enlisted to take the great-grandmother back to her home after a gala birthday celebration in her honor held at an uncle's house in the suburbs. But the late-night car ride is seemingly endless, punctuated by the narrator's clumsy attempts at conversation with this very private, very elderly woman whom he barely knows. During periods of awkward silence, he mentally contrasts Los Angeles to the place he used to live:

> I felt like we were underground, not on top of the earth, that I was driving, but not in control of the destination we were rolling toward. I'd had rides like this on roads away from the city back home. I'd driven this very car, headlights turned off, the windows down, wind blowing, and I'd felt a similar sensation, but there it was the stars and the moonlight on the cactus and weeds and on the dirt and rocks. That felt old to me, like I was seeing the past, seeing it like every man who ever passed through at any time saw it. But freeway driving was the future, with its light hanging over us in squares and rectangles, circles and ovals, their reflections on the waxed colors, on steel and aluminum, on the billboards of young women rubbing themselves with tanning oil, or young men with rippling stomach muscles modeling pants. (69)

The narrator clings to the stability of this distinction between Los Angeles and his former home, even as he acknowledges similarities between them and his own recurring feelings of powerlessness in both places. For him the contrasts are a consoling narrative of progress. His home and Los Angeles are polar opposites—one is traditional, the other is modern; one is natural, rugged, and unchanging; while the other is glossy, commercial, and dynamic. But this conception also poses problems for him, in the sense that it separates two ideals that he would like to uphold. On the one hand, he desires to be part of a traditional close-knit family, and on the other, he is captivated by Hollywood and its promise of glamour, social status, and membership in that larger, exclusive community imagined by movie culture. At the age of sixteen, the narrator's own sister had been inspired by the great-grandmother's example to run away to Hollywood, and now he, half-heartedly, has followed suit. Indeed, most members of his extended family have opted for an atomized existence. They are scattered throughout Mexico and the Southwest and see one another only rarely, to the degree that even his mother must concede, dispersion and distance themselves are family traditions (68).

The narrator's connection to his great-grandmother holds the key to resolving his apparently incommensurable desires for family and fame. The climax of the story occurs when they finally arrive at her modest, run-down, home in Hollywood. After what seems an eternity of fumbling with the lock on the door, he helps her into the house and is greeted by an overwhelming stench of cat piss. Suddenly his attention is drawn elsewhere:

> I completely forgot about the odor while I was looking at the pictures on the walls. Hollywood photos, black and whites, of her and some other actors and actresses I didn't recognize. She was in costume in most of them, Spanish costume, but also in one which I didn't know the nationality. I felt this swirl of relief because, as much as I wanted it to be true, I always doubted, worried it was a story I'd been fool enough to believe. But now I was here, in this otherwise normal, otherwise sadly furnished, smelly house. (73–74)

There is a bittersweet quality to the narrator's relief at seeing these photographs. They are, after all, portraits of a woman whose movie career is unknown even to her family, and whose costumes suggest that she was probably typecast as a multi-purpose "ethnic woman" throughout her career. The photos' guarantee of social status is also undercut by the state of borderline poverty and isolation in which the great-grandmother presently lives. Even the narrator must admit that the famed address, with its gang graffiti and cars on blocks, belongs to a "normal house," "not much different than the ones at home really, though there was something about the street and the lawns, the palm trees all around" (71). These details, in turn, threaten his narrative of progress. The opportunities for him in Hollywood, in fact, are not much different from the ones in the place he left behind. His new job is at a Firestone station, and his sister, who used to send home sexy photos of herself dressed for a night out at the clubs on Sunset Boulevard, is now a single mother who has hired a Mexican woman to care for her child while she works.

Gilb's story pointedly illustrates the race-, gender-, and class-based exclusions that are the underside of Hollywood's glamour. The contrast between the great-grandmother's photographs and her living conditions makes a direct connection between stereotyping at the level of narrative content and restricted hiring practices at the heart of industrial production, which in turn perpetuate hierarchical divisions of labor and urban space. Returning to the question of the perceived lack of "localist" border cinema that I described at the outset of this essay, "The Magic of Blood" suggests that the solution is not so simple. It does not rest in *fronterizos'* greater proximity to Hollywood, nor in the elusive quest for an adequately mimetic representation of the border (as though Hollywood could even "accurately" represent itself). Rather, one senses, it would require substantial involvement in the production process on the part of populations who have traditionally been excluded from it. "The Magic of Blood" is not sanguine about the likelihood of this possibility. For the protagonist, the movies are but one manifestation of a pervasive consumer culture, which promotes desires for commodities, bodies, and living standards that are simply untenable for the majority of the city's residents.[27] Los Angeles is for him the set of "some chic American movie" (63), which he observes and experiences with a vague sense of detachment.

At the same time that it demystifies Hollywood, however, the story also tarnishes the nostalgic view of a happier, simpler life in the place that the narrator left behind, by suggesting that this was not the case for everyone. This critique once again arises through contradictions between what the narrator says, generally, and what he chooses to believe, in this case regarding women's roles in the household and the public sphere. His ideal of a close-knit matriarchal family, associated with his mother and his former life in the Southwest, ultimately troubles his cross-gender identification with his great-grandmother. The latter matriarch, it would appear, is well-suited to be the foundational figure of a transnational, diasporic family. The narrator clearly admires her and understands her motives for leaving Mexico to be the same as his own: "She wasn't stupid either, and she wanted all the life that most people want but can't have" (65). But his familial ideal depends upon the labor of a stay-at-home woman, like his mother, who does the work of childrearing and maintaining kinship relations. In this respect, his great-grandmother's abandonment of home and family threatens this ideal, and he transfers his negative feelings about her to his sister, whom he secretly wishes to punish for having left the family, even though he followed right behind her (64). One senses, thus, that other narratives remain untold in this story, and that from the perspective of one of the female characters, the divisions between home and Los Angeles, and the private and public spheres, might look very different.

Conclusion

The political urgency that marks Gina Valdés's manichaean portrayal of spectatorship in "Llegaba el Domingo" contrasts with a more conciliatory attitude of selective assimilation toward the mass media in the three narratives I have discussed. The writings by Trujillo Muñoz, Cantú, and Gilb are characterized less by authorial competition with the movies than they are by a sense of local authority to supplement that which remains unrepresented in mainstream depictions of border life, and allusions to the multiple "other stories" that arise through the local receptions of mass media.[28] Each story builds tension around the truth value of images: the prostitute's interior monologue that belies her actions in "Border Hotel"; the disjunction between word and image in *Canícula;* and, the glossy photos of the sister and great-grandmother in "The Magic of Blood" that are no proof of having "made it."

As a counterhegemonic strategy from the periphery, literary response to visual mass media has gained significant recognition for border authors in wider networks of distribution.[29] But this remapping of cultural production comes at the cost of replacing relatively accessible visual mass media with the more limited circulation of print media. Despite these authors' strong commitment to local communities, their writing reaches a smaller audience than movies and television, and it is marked by an increased orientation toward elite, educated readers. Another irony of these authors' success lies in the fact that their own self-presentation as *fronterizos* becomes blurred as their works are absorbed into the marketing strategies and monolingual orientations of their respective national publishing industries, which have further categorized them according to subgroups such as Baja Californian literature, Chicana/o literature, *tejana/o* literature, and proletarian literature. Here one might speak of a "reverse star vehicle" where local audiences are in a position to look beyond the authorial personae created by the publishing industries to recognize the contributions that these authors have made as cultural workers in the border region—Cantú through her ethnographic research and her long tenure at Texas A& M International University in Laredo,[30] Trujillo Muñoz through his teaching and editorial work at the Universidad Autónoma de Baja California in Mexicali, and Gilb, who cites fellow carpenters and union members as his favorite fan base (Ferguson 11).

Despite the differences in perspective among these three works, when read together they foreground key categories of analysis, such as bilingualism, that might serve as a point of departure for the study of binational film culture in the border region. They also provide a powerful critique of productions about the border that emanate from outside of the region. Above all, I would like to underscore two key aspects of this critique: First, by treating cinema in an expanded field that includes local forms of cultural expression, and by writing from a peripheral perspective vis-à-vis the Hollywood and Mexico City–based cultural industries, these narratives suggest that the production of "others" at the level of content is implicated in the division of labor of industrial production. This insight is inspired by Nancy Fraser's attempt to theorize a concept of social justice that would integrate demands for redistribution and recognition. It seems that a necessary first step in understanding the mechanisms of what she calls "bivalent oppression" (31), that is the experience of scarcity with regard to resources and social respect, would be to examine issues of cultural production and consumption in the same analytical frame, as the three authors whom I discuss in this essay have done. The relevance of this point is heightened all the more in the post–NAFTA era, given the increasing exportation of Hollywood film productions to Canada and Mexico (Bacon). Second, the three authors insist on the historical authority of lived experience at the same time that they refuse to assign discrete moments of origin and ending to mainstream discourses about the border. In all three stories, linear chronology is troubled by the persistence of memory, intertextual references, and recurring events and tropes, to the extent that one senses, narrative formulae may be altered, but never stop circulating entirely. The lingering effects of an older binational configuration of the mass media on several generations of *fronterizos* is important to bear in mind when one is considering the impact of newer forms of transnational Latina/o media in the border region that Arlene Dávila describes in her essay in this volume, such as *rock en español* or pan-American *telenovelas.* Paradoxically, it is a regional approach to border literature that reveals the continued importance of the national and the (neo-)colonial to Latina/o cultural producers working on both sides of the border.

The task of an ethnography of spectatorship is to describe what is rather than what could be or should be (Ang 149). Reading border literature about mass media, one senses the tremendous need for further empirical research on reception simply in order to broach the question of "what is" (not to mention the sheer magnitude of such an undertaking). Many historical factors about spectatorship on the border, such as the presence of live translators at screenings into the sound era, the persistence of outdoor "rural cinemas," consisting of a sheet strung between two trees, in Mexican border states through the 1970s (Trujillo *Imágenes* 64), and the relatively rapid saturation of videocassette recorders and satellite dishes in the 1980s and 1990s (Valenzuela Arce 313–20), make the border region an important counterpoint to contemporary research about cinema and modernity that takes large U.S. cities or Mexico City as its points of reference. But at the same time that I refrain from attempting to define whatever utopia may be incarnated through individual acts of consumption, I refuse to let go of the utopian impulse (the "could be") as an important part of cultural production. The works I have discussed make interventions in the cultural sphere that affirm their own status as literature at the same time that they express a keen awareness of potentially more inclusive configurations for the cultural industries.

Notes

1. I would like to thank Manishita Dass, with whom I had many engaging conversations about spectatorship while I was preparing this essay. This essay is the revised version of an earlier article, which appeared in *Studies in Twentieth Century Literature.* 25.1. (Winter 2001): 15–45.

2. The poem was published in bilingual format. The inclusion of the English translation helps to interpret some of the ambiguities in Spanish, such as the word "*casados,*" which could refer to either "married men" or to "married people."
3. A substantial body of scholarly literature on "border cinema" has developed around representations of the border generated by the Hollywood and Mexico City–based industries (see for example, Cortés; Maciel; Iglesias Prieto; Trujillo Muñoz, *Imágenes*). Among these sources, Iglesias Prieto's historical periodization of Mexican border cinema is useful for signaling moments of departure from stereotypical portrayals (e.g., Tin Tan's popularity in the 1940s; the commitment that some Mexico City directors and producers have developed toward the border region).
4. I selected de Certeau to represent this pole of argumentation about consumption because of his presence in the border region, through his Visiting Professorship in French and Comparative Literature at UC, San Diego (1978–84), and his critical writings on Latin American alterity, such as *Heterologies.*
5. The turn toward empirical audience research and reception studies follows the prominence of Althusserian and psychoanalytic theoretical models in the 1970s and 1980s. See Mayne for a general overview of these approaches.
6. On the problems of conceptualizing "resistant" or "critical" versus "passive" spectatorship, see Mayne 138, 171–72 and Ang 179, as well as their respective critiques of Janice Radway's *Reading the Romance* (Ang 98–108; Mayne 82–86). See also the responses to Lawrence Levine by Robin D. G. Kelly and T. J. Jackson Lears in the *American Historical Review.* My own discomfort about "resistant" spectatorship also stems from the researcher's assumed, yet often unarticulated, horizon of concrete expectations against which the spectator's "utopian" desire is measured, be it a functionalist coping mechanism or revolutionary activity.
7. I have been especially influenced by Michael Denning's reflections on the theoretical underpinnings of his 1987 work, *Mechanic Accents* in the "Afterword" to the 1998 edition. He writes, "If the study of readers and reading [or, in this case, viewers and viewing] is not to fall into an antiquarian empiricism, we must remember that our goal is cultural history and cultural criticism, not a history of reading. What *is* historically significant, what must be a part of any serious cultural studies, is the exploration of the ways in which audiences are organized and mobilized, how cultural movements, subcultures, and cultural institutions attempt to promote and shape readings" (263).
8. My criticism of tendencies to identify resistance with consumption coincides with Manuel Martinez's critique of what he calls "poststructural/postcolonial borderlands criticism" in his essay in this volume.
9. In border fiction, a trope related to this one is the appearance of the border itself as a theatrical space where conflictive cinematic encounters take place between nationally coded icons. Harry Polkinhorn and Tomás Di Bella Martínez's collaborative short story, "Wayne/Infante (Conversación póstuma

entre dos estrellas)" ("Wayne/Infante: Posthumous Conversation between Two Stars"), has these two cowboy stars, who never met in real life, converse in heaven about the cinematically induced fantasies that Mexicans and Anglos project onto one another's countries. Gina Valdés's poem, "The Border," and Carlos Fuentes's short story, "Malintzín de las Maquilas," perform similar operations.

10. Mulvey would go on to refine her initial theoretical observations in *Visual and Other Pleasures* (1989).
11. My description of the movie theater as an institution of the public sphere is inspired by the contributions of Hansen and Fraser on this subject.
12. See Fox on grassroots film and video (41–68) and Trujillo Muñoz on cineclub, art film, and documentary production (*Imágenes,* 75–96). One indication of the unevenness regarding production "in" versus production "on" the border is the fact that the "First Encounter of Film and Video on the Border" held in Tijuana in 1985 included no work by local filmmakers. For the next such event held two years later, an effort was made to address this problem, but this time around, the only examples of local production that the festival organizers were able to find were short promotional videos for the tourism industry, documentaries produced by academics, and video and super-eight works produced by students and *cineclubistas* (Trujillo Muñoz, *Imágenes,* 91).
13. Rivera describes how a sector of young Mexican filmmakers has also turned to the *videohome* market as a means of breaking into the Mexican and U.S. film industries, as in the case of the sleeper hit *El camino largo a Tijuana* (1987, dir. Luis Estrada).
14. U.S. studio employees also used *Titanic* as a case in point to protest the growing trend toward outsourcing national film production to Canada and Mexico (Bacon).
15. The two projects included in the exhibition were Alan Sekula's photographic suite entitled, "Dead Letter Office" at the Centro Cultural Tijuana and Revolucionarte's "Popotla-The Wall" located in Popotla, a fishing village in Baja California Norte, near Tijuana (Yard, 28–37; 180–81).
16. Recently within film studies there has been a growing trend of opposition to traditional studies of "national cinemas" that have developed around textual analyses of canonical nationally produced movies. The newer models propose that cinema be studied in an expanded cultural and historical field. They call for an overall analysis of the cinema's conditions of production and exhibition in a given country; its relation to other media and forms of cultural expression, and cultural policy; and, its relationship to the social formation (see for example, Miller; Higson). Writing about methods of historical inquiry into film reception, Klinger makes a similar argument in favor of tracing cinema's inscriptions within hypothetically interminable diachronic and synchronic fields. I welcome both of these approaches for their ability to interface with transnational media, and I find them especially appropriate for the study of film culture in countries lacking strong bases of

film production. They do require further definition for the study of film in a binational region—namely, there is no self-evident spatial or social boundary to limit a study of the border, and any definition of the region necessarily privileges certain categories of analysis over others. Additionally, cultural production from the border region may also be recruited under a number of competing categories that deny regional identification (e.g., national, diasporic, or ethnic identities). My own view of the border in this essay follows that outlined in my book: while delimiting the border according to both geographical (Herzog) and social factors (Vila; Martínez), I privilege an urban frame of reference in my analysis.

17. A case in point is that of Don Francisco Bernal, Baja California's most prolific film critic, who composed his forty-year output entirely in sonnets and rhymed *décimas* (poetic verses in Spanish consisting of ten lines, each possessing eight syllables) (Trujillo Muñoz, *Imágenes* 70–73).
18. Valentino's wedding took place one year after the first movie theater in the region appeared in Brawley, California (Trujillo Muñoz, *Imágenes* 47).
19. See for example, *The Long Good-Bye* by Chandler, "The Golden Horseshoe" by Hammett, and *Los motivos de Caín* by Revueltas.
20. Hansen's chapters on Valentino recount many of the rumors surrounding his sexuality and relations with women, for example (245–294).
21. La Malinche, also known as Malintzín Tenepal or Doña Marina, was a young woman of Aztec origin who was offered as a slave to Hernán Cortés by an indigenous leader of the Tabasco region of Mexico. La Malinche went on to serve as Cortés's translator, adviser, and guide in dealings with the Aztecs; she also bore him a son, Martín. In Mexican history and culture, she is a figure associated with national betrayal, although recent contributions by Mexican and Chicana feminists have done much to challenge and transform La Malinche's negative image.
22. Trujillo's *Mezquite Road* and "Lucky Strike" both feature solitary male detectives as protagonists.
23. I am relying on Gunning's overall characterization of early cinema in "The Cinema of Attraction" and on Hansen's account of the historical shift in film narrative in Part I of her book (23–125).
24. Other examples from border literature suggest a similar way of participating in movie culture. See, for example, the short stories by Cisneros and Paredes cited in the bibliography.
25. This particular episode appears to be based on a real event from Cantú's life (McCracken35).
26. An interesting text for comparison to this one is Estela Portillo Trambley's comic short story "La Yonfantayn" about a Chicana's attempt to emulate U.S. movie stars and the plots of romance movies in order to seduce the Mexican man whom she loves. Her success depends precisely upon the imperfection of her imitations.
27. I am reminded here of Lila Abu-Lughod's observations in her ethnographic research about Sa'idi (Upper Egyptian) spectators of television soap operas:

"The 'uneducated public' at whom these serials are directed participates in the more common form of modernity in the post-colonial world: the modernity of poverty, consumer desires, underemployment, ill health, and religious nationalism" (207).

28. The attitudes of these writers toward the cinema is thus somewhat softer and more comfortable than Jean Franco's characterization of how individual Latin American writers associated with the Boom reacted to new found competition from the figure of the mass media superstar. Writing of Carlos Fuentes and Mario Vargas Llosa, she observes: "Both writers react to mass culture not by adopting a modernist aesthetic (despite Fuentes's trendy experiments) but rather by straining older forms of narrative to accommodate the displacement of significance from author to star" (160).
29. Trujillo Muñoz is the recipient of numerous literary prizes, among them the state prizes for novel and poetry, the national Abigael Bohórquez Essay Prize (1998), the binational Pellicer-Frost Poetry Award (1996), and the international Border Excellence award (1998); Cantú received the Premio Aztlán for *Canícula;* and, Gilb received the Whiting Award, PEN/Hemingway, and PEN/Faulkner awards for *The Magic of Blood* (Rodríguez).
30. Norma Cantú recently moved to the University of Texas, San Antonio, after many years at Texas A & M International University, Laredo.

Works Cited

Abu-Lughod, Lila. "The Objects of Soap Opera: Egyptian Television and the Cultural Politics of Modernity," *Worlds Apart: Modernity Through the Prism of the Local.* Ed. Daniel Miller. London: Routledge, 1995: 190–209.

Ang, Ien. *Living Room Wars: Rethinking Media Audiences for a Postmodern World.* New York: Routledge, 1996.

Bacon, David. "Is Free Trade Making Hollywood a Rustbelt?" *LA Weekly* (November 23, 1999).

Cantú, Norma Elia. *Canícula: Snapshots of a Girlhood en la Frontera.* Albuquerque, NM: University of New Mexico Press, 1995.

Cisneros, Sandra. "Mexican Movies," *Women Hollering Creek and Other Stories.* New York: Random House, 1991: 12–13.

Cortés, Carlos E. "International Borders in American Films," *Beyond the Stars.* v. 4 Bowling Green, OH: University of Ohio, Bowling Green, 1993: 37–49.

de Certeau, Michel. *Heterologies: Discourse on the Other.* Trans. Brian Massumi. Minneapolis, MN: University of Minnesota Press, 1986.

———. *The Practice of Everyday Life.* Trans. Steven Randall. Berkeley, CA: University of California Press, 1984.

Denning, Michael. *Mechanic Accents: Dime Novels and Working-Class Culture in America,* Revised Ed. London: Verso, 1998.

Di Bella Martínez, Juan Antonio. "Wayne/Infante (Conversación póstuma entre dos estrellas) (Wayne/Infante: Posthumous Conversation between Two Stars),"

Encuentro Internacional de literatura de la frontera. Borderlands Literature: Towards an Integrated Perspective. Eds. Harry Polkinhorn and José Manuel Di Bella. Mexicali/Calexico: Ayuntamiento de Mexicali and Institute for Regional Studies of the Californias, San Diego State University, 1990: 68–84.

Espinosa, Paul, dir. and prod. *The Border* (television program) broadcast on PBS, September 1999.

Fox, Claire F. *The Fence and the River: Culture and Politics at the U.S.-Mexico Border.* Minneapolis, MN: University of Minnesota Press, 1999.

Ferguson, Kelly. "Union Spotlight: Profiles of Union Members in Action," *Union Plus* 4.3 (Summer 1995): 10–11.

Franco, Jean. "Narrator, Author, Superstar: Latin American Narrative in the Age of Mass Culture," *Critical Passions.* Eds. Mary Louise Pratt and Kathleen Newman. Durham, NC: Duke University Press, 1999.

Fraser, Nancy. "Rethinking the Public Sphere: A Contribution to the Critique of Actually Existing Democracy," *Habermas and the Public Sphere.* Ed. Craig Calhoun Cambridge, MA: MIT Press, 1992: 109–143.

———. "Social Justice in the Age of Identity Politics: Redistribution, Recognition, and Participation," *Culture and Economy after the Cultural Turn,* Eds. Larry Ray and Andrew Sayer. London: Sage, 1999: 25–52.

Fuentes, Carlos. "How I Started to Write," *Myself with Others: Selected Essays.* New York: Farrar, Straus, Giroux, 1988: 3–27.

—. "Malintzín de las Maquilas (Malitzin of the Maquiladoras)," *La frontera de cristal (The Crystal Border).* Mexico City: Alfaguara, 1995: 129–60.

Gilb, Dagoberto. "The Magic of Blood," *The Magic of Blood.* New York: Grove, 1993: 63–74.

Girven, Tim. "Hollywood's Heterotopia: U.S. Cinema, the Mexican Border and the Making of Tijuana," *Travesia: Journal of Latin American Cultural Studies* 3:1–2 (1994): 93–133.

Gunning, Tom. "The Cinema of Attraction: Early Film, Its Spectator, and the Avant-Garde," *Wide Angle* 8.3–4 (1986): 63–70.

Hammett, Dashiell. "The Golden Horseshoe," *The Continental Op.* New York: Vintage/Black Lizard, 1992: 43–90.

———. *The Long Good-Bye.* Boston: Houghton, 1953.

Hansen, Miriam. *Babel and Babylon: Spectatorship in American Silent Film.* Cambridge, MA: Harvard University Press, 1991.

Herzog, Lawrence. *Where North Meets South: Cities, Space and Politics on the U.S.-Mexico Border.* Austin, TX: Center for Mexican American Studies, University of Texas at Austin, 1990.

Higson, Andrew. "The Concept of National Cinema," *Screen* 30.4 (Autumn 1989): 36–46.

Iglesias, Prieto, Norma. *Entre yerba, polvo y plomo: lo fronterizo visto por el cine Mexicano* (*Amidst Grass, Dust, and Lead: The Border As Seen by the Mexican Cinema*) 2 vols. Tijuana: COLEF, 1991.

Kelly, Robin D. G. "Notes on Deconstructing 'The Folk,'" *American Historical Review* 97.5 (December 1992): 1400–1408.

Klinger, Barbara. "Film History Terminable and Interminable: Recovering the Past in Reception Studies," *Screen* 38.2 (Summer 1997): 107–128.

Lears, T. J. Jackson. "Making Fun of Popular Culture," *American Historical Review* 97.5 (December 1992): 1417–1426.

Maciel, David. *El Norte: the U.S.-Mexican Border in Contemporary Cinema.* Border Studies Series 3. San Diego, CA: Institute for Regional Studies of the Californias, San Diego State University, 1990.

Martínez, Oscar J. *Border People: Life and Society in the U.S.-Mexico Borderlands.* Tucson, AZ: University of Arizona Press, 1994.

Mayne, Judith. *Cinema and Spectatorship.* New York: Routledge, 1993.

McCracken, Ellen. "Norma Elia Cantú," *Dictionary of Literary Biography.* v. 209: *Chicano Authors, Third Series.* Eds. Francisco A. Lomelí and Carl R. Shirley. Detroit, MI: The Gale Group, 1999: 34–39.

Miller, Toby. "Screening the Nation: Rethinking Options," *Cinema Journal* 38.4 (Summer 1999): 93–97.

Mulvey, Laura. "Visual Pleasure and Narrative Cinema," *Screen* 16.3 (Autumn 1975): 6–18.

———. *Visual and Other Pleasures.* Bloomington, IN: Indiana University Press, 1989.

Noriega, Chon. "Flimflam Film List," *UCLA Magazine* (Fall 1998): 72.

Paredes, Américo. "The American Dish," *The Hammon and the Beans and Other Stories* Houston, TX: Arte Público Press, 1994: 205–230.

Please Don't Bury Me Alive! Dir. Efraín Gutiérrez. San Antonio, Texas, 1976.

Portillo Trambley, Estela. "La Yonfantayn," *Rain of Scorpions and Other Stories.* 2nd ed. Tucson, AZ: Bilingual Press, 1993: 99–110.

Proffitt, T. D., III. *Tijuana: The History of a Mexican Metropolis.* San Diego, CA: San Diego State University Press, 1994.

Revueltas, José. *Los motivos del Caín.* Mexico City: Era, 1979

Rivera J., Héctor. "Los nuevos cineastas mexicanos se fugan a Estados Unidos, (The New Mexican Filmmakers Flee to the United States)," *Proceso* 850 (February 15, 1993): 48–51.

Rodríguez, Alvaro. "The Magic of Gilb," *The Austin Chronicle* 14.18 (January 6, 1995): 24.

Sánchez, Rosaura. "Gina Valdés," *Dictionary of Literary Biography* v. 122. *Chicano Authors, Second Series.* Eds. Francisco A. Lomelí and Carl R. Shirley. Detroit MI: Gale Research, 1992: 277–280.

Shohat, Ella and Robert Stam. *Unthinking Eurocentrism: Multiculturalism and the Media.* London: Routledge, 1994.

Titanic. Dir. James Cameron. Twentieth Century Fox and Paramount Pictures, 1997.

Trujillo Muñoz, Gabriel. "Border Hotel," *Fiction International* 25 (1994): 159–165.

———. "Hotel Frontera," (unpublished manuscript), n.d.

———. *Imágenes de plata: el cine en Baja California.* (*Silver Images: The Cinema in Baja California*). Tijuana: XV Ayuntamiento, El Instituto de Cultura de Baja California, la Universidad Autonóma de Baja California, and Editorial Larva, 1997.

———. "Lucky Strike," (unpublished videoscript), May 1994.
———. *Mezquite Road.* Mexico City: Planeta, 1995.
Tsivian, Yuri. *Early Cinema in Russia and Its Cultural Reception.* Trans. Alan Bodger. New York: Routledge, 1994.
Valdés, Gina. "The Border," *Encuentro Internacional de literatura de la frontera. Borderlands Literature: Towards an Integrated Perspective.* Eds. Harry Polkinhorn and José Manuel Di Bella. Mexicali/Calexico: Ayuntamiento de Mexicali and Institute for Regional Studies of the Californias, San Diego State University, 1990: 321–322.
———. "Llegaba el Domingo/Sunday Arrived," *Comiendo lumbre/Eating Fire.* Colorado Springs, CO: Maize Press, 1986: 10–11.
Valenzuela Arce, José Manuel. "Tijuana: la recepción audiovisual en la frontera," *Los nuevos espectadores: cine, televisión y video en México.* (*The New Spectators: Cinema, Television, and Video in Mexico*). Ed. Néstor García Canclini. Mexico City: IMCINE, 1993: 298–329.
Vila, Pablo. *Everyday Life, Culture, and Identity on the Mexican-American Border.* Ph.D. diss., University of Texas, Austin, 1994.
Yard, Sally, ed. *inSITE 97: Private Time in Public Space.* San Diego, CA: Installation Gallery, 1998.

Mapping Latinidad

Language and Culture in the Spanish TV Battlefront

Arlene Dávila

> After reading your cover story "Must Sí TV" by Elia Esparza, I couldn't help but think—it's about time! As a Mexican American and a Tejano, I have always been disappointed with television's lack of recognition of my culture. Since most TV stations are based in Miami, Mexico City, New York, or L.A., the views expressed or characters portrayed on most shows have been based on those cities' demographics.
>
> —John Barraza, Houston, Texas

> It's great that networks like Galavision are heading in a new direction as far as bilingual programming goes. But at what cost? When I was younger, I lost my interest in my roots and my second language, which is Spanish. I became too assimilated. I realized that relearning Spanish would benefit me in my job. I had to do something. I didn't have the money to go back to college, so I watched a lot of Spanish television—no bilingual television—and read books. I learned how to speak Spanish better. I believe bilingual television will cause the younger generation to be cut off from their roots.
>
> —Beatriz Montelongo, Lubbock, Texas

The letters above were sent to the editor of *Hispanic Magazine* in reaction to an earlier article announcing the development of "Must Sí TV," the first television programming service targeting bilingual and English-dominant Latinos. Both letters evidence how closely intertwined specific media developments are with the very heated language debate among Latinos. The question of whether Spanish is integral to Latino identity is one that the Spanish-language TV networks have historically made central to their own marketing by positing Spanish—and accordingly themselves—as primary conduits of Latinidad, although not without controversy and criticism. The first reader's disappointment at the Spanish TV networks' lack of representation of his culture and the second's concern that she would be too assimilated if she did not speak Spanish evidence the kind of dilemmas that arise from the strict association of the Spanish language with U.S. Latinidad.[1]

This article examines the politicization and general treatment of language by the Spanish TV networks through an analysis of their philosophy and recent changes in programming and what that suggests about the imaging and conceptualization of U.S. Latinos. My goal is to explore what these media developments suggest about the place of language in the imagining and conceptualization of U.S. Latinidad as well as about the transnational dimensions of this identity. Specifically, I want to emphasize the nationalist underpinnings that underlie the network's treatment of language and their definition of "Latinidad," and how it affects both the political value and inclusiveness of Latino/a as an identity category, and the ability of culture-specific media to provide alternatives to dominant nationalist frameworks that strictly equate a "people" with one culture and one language. In so doing, this article contributes to the overall concerns of this volume by questioning prevailing discourses that posit transnationality as an alternative to hegemonic versions of U.S. nationalism and that are, in large part, indebted to pioneering borderlands work on the transnational implications of Chicana/o identities at the U.S.-Mexico border.

My discussion is based on recent media developments involving the growth of new bilingual shows targeting the English-dominant Latino/a, such as Must Sí TV's *Funny is Funny* and *Cafe Olée* as well as a rapid growth in bilingual and English-dominant print media that challenge traditional definitions of Hispanics as a Spanish monolingual constituency. Such notions have long dominated Hispanic media. But we now see an increase in the use of "Latino/a" as the new media's main form of address for this imaginary constituency, rather than "Hispanic"—long used to mark the importance of the Spanish language as the primary identity marker—as well as a growing interest in the bilingual, English-dominant, urban Afro-Latino or the "home girl" as the prototype of the Latino/a. Many recent publications have even adopted mission statements evoking Latino empowerment and self-representation, announcing that their writers and staff have indeed "lived and experienced" Latino culture and are thus able to represent "all sides of our story." This trend evidences the emphasis placed on the issue of representativity in these media's marketing and self-presentation.

Yet nothing has provoked more excitement—and fear—in the Hispanic media and marketing industry in recent years than the sale and revamping of Telemundo, the number-two rated Hispanic network, after Sony Pictures Entertainment bought it in 1998 in conjunction with Liberty Media Corporation, Apollo Investment Fund III, and Bastion Capital Fund. This event marked the inclusive acquisition of a Spanish network by a group of global corporations, and the entry of one of the major mainstream entertainment companies (Sony) into the U.S. Hispanic market, leading to a major revamping of the station's programming and bringing to the forefront questions of purpose and intention on the part of the "American" investors. Of concern to the Hispanic advertising community was whether this sale would culminate in Telemundo's "Anglicization" and the ultimate eradication of this Latino/Hispanic network space. News about the acquisition repeatedly pointed to the fact that Telemundo's new president and CEO were unfamiliar with the Latino community and that Latinos lacked representation in its management team, which prompted questions about the network's legitimacy and authenticity (Esparza; Mejia). A further cause of excitement and distress were public announcements about Telemundo's plans to reshape Hispanic television through original programming. Although limited in scope, as I shall argue below, such initiatives are opening spaces for alternative discourses of Latinidad, with different media outlets contesting the dominant Spanish language-centered definition of Hispanics/Latinos in order to carve out a specific market niche that is clearly distinct from their competitors.

I will argue that the very location of Latinidad is at stake in these developments. While the Spanish language and the importation of Latin American programming are at the center in the transnationalization of Latinidad beyond U.S. borders, English and U.S.-made initiatives function as an additional venue for consolidating the United States as the primary terrain of Spanish-language TV. Additionally, I will suggest that, whether departing from the discursive realms of Latin America or the United States, both models are similarly limited by their equation of language and Latino representativity in ways that subsume other differences among Latinos and are, moreover, confined by the exigencies of dominant notions of U.S. citizenship. While citizenship is certainly a contentious term, my emphasis will be on assessing the extent to which assertions of cultural differences intersect with dominant norms of American citizenship that give preeminence to white, monolingual, middle class producers of and contributors to a political body defined in national terms. That is, my concern is not with citizenship as a neutral category or with the U.S. as an apolitical body similarly devoid of cultural meaning, but with how notions of citizenship, belonging and entitlement are directly intertwined and predicated on dominant U.S.

nationalist categories. Such categories conflate race, culture, and language with "nationality," establishing the hierarchies and coordinates against which cultural and linguistic differences are ultimately evaluated (Ong; Williams). It is therefore these hierarchies that frame media discourses of Latinidad and the media's treatment of language in ways that potentially communicate to and about Latinos the terms of their claim to belonging within the political community of the United States. I start by analyzing the fare and philosophy of Univision. This number one rated network with a claimed share of 80 percent of the Latino/a viewership is the primary promoter of Latinidad as what Arjun Appadurai has called an "ethnoscape," a diasporic community transcending the U.S. and Latin American nation states. I then turn to analyze the U.S.-based approaches to Latinidad presented by Telemundo under its new management, and the predicaments involved in both of these modes of representation.[2]

Univision and Telemundo: Towards One Vision/One Culture

Since its establishment as SIN in 1961, Univision has been the major disseminator of a Spanish- speaking version of Latinidad targeting one segment of the highly diverse U.S. Latino population—the recently immigrated and monolingual Spanish-speaking Latino. While U.S. Latinos differ in terms of levels of acculturation and language use, it is this model viewer's "unique" need for culturally and language specific programming that has historically provided Univision's marketing edge and rationale for existence as a Spanish monolingual network.[3] Emphasis on this audience has also sustained the network's "transnational" orientation and linkages with the Latin American media market. Specifically, while Univision has historically sold itself as the station for U.S. Latinos—touted as "El Canal de la Hispanidad, una herencia, un idioma, un canal" (The Channel of Hispanidad, one heritage, one language, one channel)—, it has largely operated as a "transnational" rather than an ethnic media by importing cheaper Latin American programming into the U.S. market instead of investing in the production of original programming.[4]

Dependency on Mexican products was most marked until the 1980s, when Mexico's Televisa provided 80 percent of its programs, although dependency is still at play in most contemporary U.S. produced Hispanic shows.[5] New patterns for importing Latin American cultural products include the importation of Latin American actors and of shows that are produced in the United States and repackaged as "U.S. Hispanic productions." Through this new strategy, Univision has continued to produce Spanish-speaking shows featuring only native speakers of Spanish and assuring in this manner the appeal and marketability of its programs throughout Latin

America. These are issues of central concern to Univision's owners, U.S. entrepreneur Jerrold Perenchio and two of the largest Latin American media empires, Mexico's Televisa and Venezuela's Venevision. They have a vested financial interest in maintaining a close language and programming synergy between the U.S. Hispanic and Latin American media markets (Wilkinson 86). Some examples of this pattern of cultural production include Don Francisco and his successful program *Sabado Gigante,* which was imported from Chile to Miami in the mid-1980s, and the recruitment of Puerto Ricans Giselle Blondet, Rafael José, and Mexicans Ana Maria Conseco and Fernando Arau to lead *Despierta America,* Univision's version of ABC's "Good Morning America."

Such close language and programming synergy between the U.S. Hispanic and Latin American media markets is additionally important in assuring a level of identification with Univision's programming among its audiences in both the United States and Latin America. Televisa has long been the most significant exporter of programming to the rest of the Spanish-speaking world, and as a result, the dominance of Mexican programming in Univision is not altogether an anomaly to Latin American consumers outside of Mexico (Wilkinson). But, as I shall argue below, this does not imply that the programming has entirely shed its Mexican identity in the eyes of U.S. Latinos or is unproblematically consumed as a pan-Latino product.

Beyond economic considerations, Univision's heavy fare of imported shows, actors, and performers responds also to its self-appointed role as the keeper of Latin American culture and as the conduit between U.S. Latinos and "their" culture. Univision thus positions itself as the primary venue in which U.S. Latinos can connect or reconnect with a world that they may or may not have experienced but that, as they are continually told, is nonetheless a representation of "Latin America" and thus of "their" heritage. Being Hispanic or Latino in the United States, according to Univision, thus means being able to recognize the latest Mexican *farandula* (celebrity), being conversant with details of Hurricane George's devastation of the Caribbean (news that is only peripherally covered in mainstream U.S. media), the most recent soccer match, the latest Miss Venezuela, the "right Spanish," or other bits of culture/knowledge through which Latinos can prove their Latin American prowess and cultural knowledge to themselves and others. The result is a transcontinental view of Latinidad, in which the latest *conjunto* band from Los Angeles is juxtaposed with the newest pop artist from Puerto Rico and in which guests from TV shows are as likely to be flown in from Panama as from New York. In this version of Latinidad, Latin America is constantly reinstated as the central signifier in the U.S. Latino landscape.

Such a Latin American–centered approach to Latinidad is a dominant trend in most contemporary representations of Latino identity that is also evidenced in other fields of cultural production. Following the nationalist underpinnings underlying contemporary representations that view cultures as bounded and contained entities, tied to a territory, a past and a heritage, it is Latin America rather than the deterritorialized U.S. Latino culture that has traditionally been valorized as the source of cultural authenticity in Latino/Hispanic culture.[6] Univision's stance is therefore not an arbitrary one, but is rather guided by existing hierarchies of representation that necessarily impact Univision's attempts to become the "representative" medium for Latinos.

At the same time, such strong connections with Latin America have generated criticism about the network's lack of attention to the needs and experiences of U.S.-based Latinos, who are marginalized if not altogether absent from these representations. In New York City, the Spanish network's Latin American-centered outlook was a repeated concern of some media activists and grassroots groups such as "La Comuna," a New York City-based not-for-profit community group, which even called for a general boycott against Univision and Telemundo in 1998. As part of their overall dismissal of U.S. domestic news in favor of Latin American stories,[7] these networks had failed to broadcast Bill Clinton's 1998 State of the Union address and the community group claimed that Hispanics were being deprived of information that is "vital for the growth and development of their community" (*El Diario* 20). Similarly, media activists in New York City have voiced concerns like those stated by the Houston reader at the chapter's opening that these stations reflect the interests of a few people who dictate from Miami what Hispanics should see and hear in the media. During my interviews with members of the New York chapter of the Hispanic Media Coalition, whose members are mostly Puerto Rican, this issue surfaced in their resentment of the dominance of Cubans in the Hispanic media and marketing industry, which they saw as an impediment to raising the visibility of Puerto Ricans. These concerns point to the public association of the media with particular Latino nationalities or particularized locations, a phenomenon that impedes the unqualified acceptance of the networks as truly pan-ethnic Latino products.

At the heart of these concerns is the fact that the transnationalization of media flows provides little space for material that fully addresses the concerns and sensibilities of U.S. Latinos. But such contentious issues remain hidden and invisible, ultimately rendering these texts assimilationist in content. What Herman Gray (1995) has discussed with regard to black-oriented TV shows is also true for Hispanic media: These media completely eliminate and marginalize cultural differences in the name of "universal similarity" among and across Latino subgroups. Obviously, I am not suggesting that U.S. Latinos could only identify with U.S.-made products created by U.S.-based Latinos. An extensive literature has already shown how highly problematic and reductionist it is to try and ascertain levels of authenticity in different media products, and such a literature has also warned against dismissing the localized meanings people find in global texts (Ang; Wilk). On the other hand, it is naive to ignore the embedded inequalities that are forged by this common Latin American media market, in particular, the segregation of U.S. Hispanics to the level of consumers rather than producers of representations of Latinidad.[8] As many critics have argued, U.S. Latinos have become doubly subordinated and invisible both in the mainstream media as well as in the Hispanic media that so openly declares that it means to represent them (Wilson and Gutiérrez). Neither of these media provide spaces for probing U.S. Latinidad by acknowledging or showcasing differences among U.S.-based Latinos and Latin Americans in ways that could potentially facilitate critical dialogue about the racialization of U.S. Latinos or about what it is to be Latino in the United States. These issues are never part of a Mexican *telenovela.*[9]

Ironically, it is Univision's rival Telemundo that, despite promising more complex representations of what it means to be Latino in this country, has most clearly articulated the dominant assimilationist view of the Hispanic TV networks. This is because, in contrast to Univision's Latin American-centered world, Telemundo has adopted a new philosophy about what it means to be Latino in the United States. Telemundo's philosophy is evident in its new slogan, "The Best of Both Worlds," which makes direct reference to Latinos' partial involvement in the United States and the Anglo world. Yet, what exactly is the "best" of these worlds that are increasingly encountering each other through Latinos, and on what basis are these traits considered the "best" of each world?

A cursory view of the promotional video Telemundo shown in 1998 during its marketing presentations to Hispanic advertising agencies and other prospective advertisers seems to hold that it is *flan,* chihuahuas, and *guayaberas* that Latinos have to offer to enrich American collies, t-shirts, and apple pie. By flashing the following equivalences: "Flan + Apple Pie = Telemundo"; "Collie + Chihuahua = Telemundo"; "Ketchup + Salsa = Telemundo"; "Santa Claus + three Kings = Telemundo," the network's 1998 advertisement touted itself as the symbiotic union of U.S. American and Latin American culture.[10] Other images are shown harmoniously blending into one another to tell the story of Latinos' involvement in two cultures, two traditions, and two languages. The Latino world shows Latinos dancing with friends and family to a band of Mexican ranchero-type musicians (whose rhythms are nonetheless more *salseado*). The words "the Journey" and "the Joy" are superimposed softly over these images as if to reinforce the feeling of enjoyment communicated by the multigenerational images of young heterosexual couples dancing, and of young children playing and dancing with an elderly grandmother figure. A second section, introduced by the words "The Struggle," "Two Traditions," "Two Cultures," "Two Languages," moves to a more urban environment and into a transitional state that is now communicated by a Mexican mural and a low-rider car. These two images dominate the scene against which young Latinos, some dark, some indigenous looking, greet friends and stride around the neighborhood.

This is a world where everyone still knows one another, a world that is sexualized yet nonetheless dominated by rules: A young Latino is shown respectfully greeting a young woman (who is accompanied by another young man) with a kiss to her hand, while openly flirting with a presumably unattached young woman. The third and last section, introduced by the words "The Dream," takes us to a more suburban environment. The opening scene depicts a little girl drinking from a water fountain in a park. This scene is quickly followed by an image of life in corporate America. Here the same Latino models, the old and the young who have traded their relaxed clothes for business suits, are now walking busily and seriously around a corporate headquarters. The active verbs "Hope," "Climb," "Reach," and "Arrive" are now shown as a soft, aspirational tune, with violins replacing the more Latin rhythmic music, stresses Latinos' successful incorporation into a "non-ethnicized" world.

I do not think that I would be reading too much into this video clip to argue that it embodies the vision of the reborn Americanized Hispanic citizen in concert with dominant values of American citizenry as well as the ghetto-to-corporation aspirational image of Latinos disseminated by the Hispanic TV networks. As if reminding Latinos, "Yes, keep your culture but keep it packaged," these images reflect back to us the hierarchy of values on which the Anglo and Latino unity is ultimately predicated. This is far from the station's claim that, as the clip states, it will provide a "world without frontiers, where your words don't need to be translated, that has the best of you on screen, that belongs to you and that you would want to give away to your children, and that is inspired by you."

A look at Telemundo's opening programming further confirms that the station's philosophy of Latino representation is far more daring than its implementation. Indeed, most of the programming under the new management consists of Spanish derivations of old American sitcoms or game shows, like *The Dating Game, The Newlyweds Game,* and *Who's the Boss.* Its variety and entertainment shows are no different from Univision's as they rely on Latin American specials like concerts and pageants. Moreover, not unlike Univision, Telemundo is also directly tied to Latin American media and hence to assuring the same linguistic programming synergy between the U.S. Hispanic and Latin American media markets. The most noted example of this trend is Telemundo's programming partnership with Televisa's biggest rival, Mexico's TV Azteca, producer of some of the network's most successful *telenovelas* such as *Mirada de Mujer.*[11]

Telemundo's original programs thus have not departed from the dominant trend of relying on imported Latin American talent and from filming in Mexico. The Latin version of *Charlie's Angels,* the failed *Angeles* (canceled after a short run), which claimed to showcase "Latina power," is a good example of this trend. The actors playing the three female detectives are from Mexico, Argentina, and Colombia, where they had artistic and modeling careers prior to their recruitment for the show, and the episodes are filmed in Tijuana by an American production company.[12] Meanwhile, another show, *Rey y Reyes,* is set in an imaginary U.S. Latino city, in "Rio Lobo," somewhere between the U.S.-Mexico border, and *Angeles* is located in the "not-so peaceful town of Costa Brava," an unspecified coastal town in the United States, markedly stripped of history or regional flavor. These are shows that are supposed to take place in the United States, yet most of the street, road, and landscape signs are in Spanish, everyone speaks Spanish, and the characters' Latinidad is stripped of any specific ethnic or national background. While the actual national backgrounds of the actors were made public in press releases and magazine articles and their national accents are perceptible in the show, they are nevertheless meant to represent "generic" Latinos whose histories and ethnic backgrounds are never developed in the show.

Replacing this original lineup, Telemundo is currently producing shows that are seeking to draw more closely from the tensions and experiences of Latinos in the United States. Among them are the 1999 *Solo in America,* which I will discuss in greater detail below, and *Los Beltranes,* a Latino version of *Archie Bunker,* which features a racist, homophobic Cuban bodega owner in Los Angeles, his Chicano son-in-law, and a gay Spaniard neighbor. As a comedy based on tensions among U.S. Latinos, these shows represent a novel development, but it remains to be seen how this trend develops in the future. For now let us consider the networks' treatment of language and what it suggests about the possibilities for expanding, or at least, complicating the public representation of Hispanidad/Latinidad in the Spanish TV networks.

Language and the Terrain of Latinidad: Toward the Best of One or Two Worlds?

The Spanish language is central to maintaining the programming synergy between the Latin American and U.S. markets. Not only is Spanish the most powerful catalyst for consolidating the unity of a transnational Spanish-speaking media community in the United States and Latin America,[13] but it is also a central political symbol that unifies U.S. Latinos and Latin Americans and is the primary vector of Latinness in the United States. It is therefore not surprising that both Univision and Telemundo have largely made Spanish central to their operations and self-presentations, treating it as an issue of "cultural citizenship." The protection of Spanish is construed as central to Latinos' struggle to maintain their culture in the U.S. as evident in the stations' emphasis on so-called "correct" and "generic" Spanish, in their abstention from Spanglish, and most of all, in their appropriation of the role of guardians and instructors of the language within the larger Hispanic community. Expressions like "tu idioma," and "nuestro idioma," "la herencia del idioma que nos une en hermandad," ("your language," "our language," "the language heritage that unites us in common brotherhood") are their common discursive devices.

Indeed, the networks' emphasis on Spanish is not ill-founded. After all, they are operating in a context where, from the standpoint of many corporate advertising clients, Spanish and culturally specific advertising and programming are not only unnecessary but also an impediment to Latinos' "assimilation" into U.S. society—their incorporation without vestiges of "tainted" culture or language. In the words of an L.A. account executive of Mexican background, referring to his previous experience as brand manager for a major pharmaceutical corporation: "From their [corporations'] standpoint, our marketing presentations confirm their suspicions that Latinos don't want to be Americans, and what they say is 'let them eat English,' 'let them be American.' In this context, the networks' historically stubborn insistence on Spanish as the conduit of Latino/Hispanic identity, their reticence against bilingual programming, or against featuring Spanglish or mixing languages also needs to be considered in relation to the stations' self-appointed role as public guardians of the language that is seen as the embodiment of Latino identity, an identity that is not only shunned, by some, but considered a threat to U.S. national imagery.

What we cannot assume, however, is that language and cultural visibility always also mean social gains or political entitlements nor can we forget that, as noted by Juan Flores and George Yúdice, the commercial use of Spanish is not about the recognition of Latinos but about constituting them as contented consumers. The question, then, is whether what the networks air necessarily provides for an unqualified notion of cultural citizenship as involving the expansion of "claims to entitlements" (Rosaldo and Flores) or whether it also represents a reformulation of the frameworks of recognition and debate, de-stabilizing pervasive constructs of citizenship, nation, and race. Specifically, we need to inquire whether what the Spanish TV networks project actually extends the coordinates around who and what is considered an American and promotes a more complex view of what it means to be Hispanic/Latino in the United States or whether, instead, it helps validate dominant norms of good American citizenship in ways that reproduce rather than challenge dominant race/class and gender norms at play in U.S. society.

Let us consider the networks' promotion of the so-called generic Spanish, what the industry calls the "Walter Cronkite Spanish" (the unaccented generic or universal Spanish), which is supposedly devoid of regionalism or a traceable accent, as the primary venue for addressing U.S. Hispanics. Like any "standard" form of language, "Walter Cronkite Spanish" is of course itself a "discursive project" that, by reproducing a particular language ideology and particular social distinctions (Wollard), privileges some accents and modes of speech over others. In fact, the Walter Cronkite Spanish has largely served as a cloak for the "mexicanization" of the Spanish language. Given that Mexican Americans constitute 65 percent of all Hispanics and that Mexican soap operas and programming dominate the U.S. Hispanic airwaves, it is the Mexican dialect, accent, and mannerisms that are generally favored as the embodiment of generic Hispanicity. While the media distribute the particular sociolect, accent, and mannerisms of mostly upper class Mexican Spanish as "representative" of the Hispanic market, Caribbean Spanish is hardly heard in generic advertisements and is highly edited. For example, the Cuban Cristina Saralegui and Puerto Rican Ray Arrieta, both highly popular Univision entertainers, have publicly revealed the pressure they have faced to tone down their accents. Cristina, who has achieved considerable influence in this industry, was able to keep her Cuban accent and have it accepted as a trademark of her TV personality, but Ray had to shed his "Ay bendito"—a common Puerto Rican expression—after the first filming of his Univision program *Lente Loco* (Crazy Lens).[14]

The stations' preoccupation with language purity has also led to the inauspicious containment of language difference among U.S. Latinos, for whom language is not only an issue of different Spanish accents but, also most critically, of levels of competence, ability, and ease with both English and Spanish. As a case in point, the stations' traditional concern over the use of "correct" Spanish as a validation of a central axis of U.S. Latino culture also functions to subjugate language and cultural difference among Latino/s, expressed in Spanglish or code-switching. As Ana Celia Zentella's study of language use among Puerto Rican children in New York notes, neither "English" nor "Spanish" can fully describe their linguistic repertoire, which can range across a wide bilingual/multidialectal spectrum. Although differentially valued or stigmatized in relation to dominant culture, this spectrum functions as a set of linguistic resources and does not invalidate the children's claim to a Puerto Rican identity. As Ana Celia Zentella and others have noted, Spanglish is at the heart of a wealth of U.S.-generated literary productions as well as at the heart of music and expressive forms, such as Latin hip-hop and rap, which are central to Latinos' experiences and identities in the United States. Yet in the networks, it is only "correct" Spanish that is reciprocally and symbiotically connected to a Hispanic/Latino identity, with the result that Spanish is used in a way that "corrects" rather than validates people's linguistic repertoires. Perhaps the best example of this strict correlation of U.S. Latinos with Spanish is provided by the promotion of the few English programs for Latinos in Univision's cable network Galavision as shows "for the Latino who also speaks English." Not only does this announcement veil the reality of code-switching and mixing between English and Spanish, but it also parts from and reinforces the very notion that all Latinos speak Spanish and that, while some "may additionally speak English," English is ancillary to their speaking Spanish.

Yet another problem with the Hispanic TV networks' preference for so-called correct language is the fact that it promotes a bounded vision of Spanish, one that contains difference and keeps it "in its place" by reinforcing distinctions among those whose background and education have given them the cultural capital of "correct speech" and those who lack it. Here it is important to note that while Spanish is indeed regarded in dominant society as a threat to English and to U.S. national integrity, it is nonetheless relatively better accepted when it is contained in grammar and in "its place" than when it is left "unbounded" as in Spanglish, code switching, and bilingualism. Accordingly, unlike the "correct Spanish," which is more likely to be considered a sign of ethnicity not at odds with U.S. dominant ideals of upward class-mobility, Spanglish is more readily associated with linguistic pollution and social disorder as the language of a "raced" underclass people (Urciuoli).

The networks have historically featured the kind of Spanish that, because it is bounded within sanctioned grammatical rules, can serve as a marker of class mobility or, as stated in the letter of one of the readers quoted at the beginning of this article, that which "would benefit people in their jobs." The Spanglish or "broken English" such as that spoken by some Puerto Rican or other Latino guests on *Cristina* is covered with a "blip" or corrected by the host's demands that they speak Spanish. I am not implying that all Latinos would like to hear Spanglish in the airwaves; in fact, while Spanglish is embraced proudly by some Latinos, it is most often generally treated with disdain.[15] The relevant and problematic issue is that concerns over language purity subordinate the status of everyday language, particularly Spanglish, as faulty speech, further aggravating the disdain among Latinos toward their languages and, by extension, toward themselves relative to the bearers of "correct speech."[16] In fact, everyday language is also derided in the few existing bilingual shows that are being touted as examples of the networks' more inclusive approach to U.S. Latinos. Telemundo originally claimed that it would reach out to the English-dominant Latino through programming,[17] but this goal has been replaced by a more conservative approach to language difference. Rather than serving as a mainstay of some programs, English appears only in the form of Spanglish, which is used selectively as a "condiment" and mostly limited to comedy shows.

In *Solo in America,* which revolves around the language and culture clash between a divorced, Venezuelan, Spanish-speaking Latina and her two bilingual and bicultural teenage daughters in Brooklyn, one of the main comedic devises is the mother's scolding of her daughters' tainted Spanish. The mother constantly insinuates the evils of becoming too Americanized. Though the youths are supposed to be fully bilingual, they are only shown speaking short phrases and mostly single words in English or Spanglish. The actual meaning of their speech, however, is almost always conveyed by the context, by an immediate translation by their mother, or else by a deliberate rephrase in Spanish. The results are constructions like "Mom, I can't believe que *tu a mi me mentiste*" (Mom, I can't believe that you lied to me) or "*Frank, no se porque te abrí la puerta, tu no entiendes* nothing" (Frank, I don't know why I let you in, you don't understand *anything*) or "Mom, have you decided, *que vas a hacer? Vas o no vas a Chicago?*" (Mom, have you decided, what you're going to do, are you going to Chicago or not?).[18] In these constructions, English words or phrases are peripheral to the sentence's meaning. This is rarely how English is used by code-switchers who are likely to incorporate whole sentences and clauses in English and switch fluently between Spanish and English as is appropriate. The program's treatment of Spanglish hence echoes what Jane Hill has identified as "mock Spanish" among Anglo speakers of English, although in this case it is the use of English among native speakers of Spanish that is marked by its dual indexicality. On the one hand, the use of English is meant to represent a symbolic connection with the acculturated Latino, but on the other, it indirectly conveys a debasement of Spanglish and its speakers. What these shows do for its main audience of native Spanish speakers is to assert that anything that is not "correct" Spanish can never convey the burden of the dialogue but only function as a comedic device.

This derisive treatment of Latinos' English and Spanglish speech is also evident in Galavision, Univision's cable network, whose mission as stated to me by a sales representative is to "provide programming alternatives for Latinos irrespective of language." Galavison has so far included only two shows for English-dominant Latinos—*Funny is Funny* and *Comedy Picante.* These shows feature Latino comedians whose material almost always pokes fun at Latino culture à la Paul Rodriguez and John Leguziamo. Even though the content of this comedic material is potentially subversive, it is still worrisome since Latinos, like blacks, have traditionally been relegated to the role of buffoons or entertainers in the mainstream media[19] and since it is mostly in comedy-style shows that English-dominant Latinos are showcased.

Ultimately, despite their claims of Latino representativity, the U.S.-based Hispanic networks have shown a limited ability to expand the range of what is accepted and promoted as "Latino" on the airwaves beyond the Spanish-dominant and trans-Latin American norm. Moreover, new programs leave unchallenged the traditional flows of media production for Latinos and do little to address the "invisible" audience of English dominant Latinos beyond mocking English and Spanglish or its Latino speakers. Nonetheless, let us for a moment consider the dissimilar tendencies that are implied in these programmatic statements about the public projection and perhaps the future conceptualization of Latino/a identities in this country. Briefly, we are faced with a growing media interconnectedness and a synergy between U.S.-based and Latin American populations, on the one hand, and, on the other, with the possibility of more U.S. based bilingual/bicultural media initiatives. The latter draw on either an expansive and widespread definition of Latino identity as transnational or else on the United States rather than Latin America as the primary reference point for Latino identity, with arguments about the need for bilingual productions functioning as a means for expanding who and what is represented as Latino on the airwaves. Yet, neither discursive proposition presents a challenge to normative ideals of U.S. cultural citizenship. The dominance of imported Latin American programming continues to limit the space for showcasing U.S. Latino sensibilities on race, identity or politics. Meanwhile, proposals for programs that represent and hence establish U.S. Latinos as an intrinsic component of U.S. American culture have fallen short of addressing the multiplicity of Latino experience in the United States in ways that reflect—rather than mock or deride—these expressions and that do not end up prioritizing dominant U.S. norms of race, language, and culture.

By prioritizing the harmonious integration of Latinos devoid of politics and difference, both of these approaches are more similar to each other than the networks' philosophies would have it. Most of all, both modes of representations reiterate the discourse of authenticity that equates language (be it Spanish or English or Spanglish) with representation, thereby subsuming race, class, and different subjectivities and backgrounds to this one issue. The result is an overemphasis on the same linguistic difference that marks Latinos as outsiders within the dominant norms of the white and monolingual U.S. national community. In the process, Spanish is reinscribed as the authentic and sole property of Latinos, never to be part of the larger, monolingual "national community."

Ultimately, any media developments regarding the representation of Latinos are likely to be more affected by the numerous interests "jumping on the bandwagon" of the Hispanic market than by the ways in which Latino communities conceptualize themselves or their identities. Simply put, the equation between language use and profit is likely to continue to affect the correlation of Latinos with Spanish, thereby impairing attempts to broaden the media's definition of Latinos or at least what they sell as "Latinos" to marketers and corporate clients. After all, Spanish is the one variable that effectively attests to the "uniqueness" of the U.S. market and that sustains the need for culture and language specific programming and advertising. For if it is not language per se that makes Latinos "Latinos," what is the need for ethnic specific media and programming in the first place? Could Latinos not just be targeted by appealing to a specific culture and lifestyle in ways that resemble market appeals to African Americans or other U.S. market segments? Or should Latinos perhaps simply be ignored as a culturally specific niche altogether? These are just some of the concerns that may explain the few innovations in the media airwaves and that are, moreover, likely to continue to impair attempts at broadening the media's definition of Latinos beyond the current dominant image of the Spanish-speaking and thus "authentic Hispanic." These culture battles in the media also hinder my ending this discussion with anything approximating a conclusion, as their outcome remains to be seen. What is certain, however, is that contestation over the Spanish networks' legitimacy and programmatic content will continue and that it is not likely to be over in the foreseeable future.

Notes

1. This article is a revised version of an essay, which appeared in the February 2000 issue of *Television & Media Studies.* For the purposes of this paper I use the term "Latinidad" to convey the definition and discourses of Latino/a identity as a pan-ethnic term encompassing every person of Latin American background in the United States irrespective of nationality, class background or race. I will also be using Hispanic/Latino interchangeably as is done by most people in the Spanish media industry although it is by the name "Hispanic" media that this industry is most commonly known. The dominance of this name is due to the business preference for the officially census-sanctioned category of "Hispanic," over "Latino," a term of self-designation more connected to social struggles and activist (Lopez and Noriega). "Hispanic" is also used to mark the importance given to the Spanish language as the key marker of Hispanic/Latino identity. Overall, however as I argue below, both terms are used equally to sell, commodify, and market populations of Latin American background in the United States.
2. Both of the networks conducted up-front presentations for prospective advertisers for the first time in 1998, pointing to their renewed efforts to market themselves and define their "uniqueness" in this new context. This section is largely based on videos and observations of these upfronts, analysis of their media kits, and interviews with people involved in their programming.

3. See Univision's "The U.S. Hispanic Market in Brief," (part of its marketing kit). This document stresses that 90 percent of U.S. Hispanics speak Spanish, that two-thirds of U.S. Hispanic adults are foreign born, and that 47.6 percent of the Hispanic population is Spanish-dominant. However, as regards the percentage of foreign born Hispanics, it is only among adults that this percentage is as high as two-thirds of the population, not among youth, which the same report claims is the fastest growing segment of the total Hispanic population. Similarly, the range of Spanish speakers subsumes the Latinos that speak English or are bilingual. The point here, however, is not Univision's self-serving manipulation of numbers, but rather that it constructs the totality of Hispanics out of one or another of its segments. Univision, makes a part—its part—into the whole.
4. I am following here the distinctions drawn by Hamid Nacify in his 1993 discussion of Iranian television. As he states, unlike Iranian exile television, Spanish media in the United States have largely been a transnational product that serves the purposes of identity maintenance and assimilation into the host society. I disagree, however, with Nacify's observation that the Spanish media have contributed to the Cubanization of Spanish TV and Hispanic society. While Cubans do dominate in Hispanic media's administration, as I mention below it is Mexican programming that has become the representation of generic "Hispanicity."
5. Until the late 1980s its precursor SIN/SICC served more as a receptacle for Mexican programming with over 90 percent of its network hours devoted to programs directly aired or imported from Mexico (Gutiérrez; Avila).
6. For more information on these issues see Rosaldo, Segal, and Handler. I also develop the Latin American basis of U.S. Latinidad in greater detail in Dávila.
7. On this issue, see America Rodríguez.
8. Relevant here is Toby Miller's suggestion for a shift of outlook from what media does to people or what people do with media to who is involved in its production and can participate in the growing transnational division of cultural labor.
9. I recognize that there have been some exceptions to this trend. The groundbreaking *Dos Mujeres y un Destino,* featuring Erik Estrada is a good example here, as is also *La Mujer de mi Vida,* featuring a young woman who moves to Miami. I would nonetheless argue that when linkages between U.S.-Latin America are developed, they almost invariably fail to problematize the tensions and problems of such connections. Scenes may be filmed in Miami but Hispanicized to such an extent that there is little or no room for recognizing the problems or experiences of U.S. Hispanics as minorities in the United States. Similarly, references to transnational U.S.-Latin American are most often employed, not from the perspective of immigrants but from that of the rich and cosmopolitan characters. These are the "floating populations" who take shopping trips to New York, go to doctors in Hous-

ton, or vacation in Miami but whose trips never compromise their sense of place and national identity.

10. Such images were part of a video produced as part of the Telemundo's new programming presentation to advertisers after SONY and its partners bought out the station. The video was presented during the 1998 semiannual meetings of the Association of Hispanic Advertising Agencies meeting in New York as well as transmitted to viewers nationwide.
11. See Barbara Belejack.
12. Press release "Telemundo's New Crime Series 'Angeles' Showcases Latina Power" released by Telemundo on January 4, 1999. The show features Patricia Manterola, Mexico's best actress of the year in 1996, Argentinean Sandra Vidal, and Magali Caicedo from Colombia. The show is filmed at Baja Norte Studies in Tijuana, Baja California. Press release written by Gabriel Reyes and Steven Chapman.
13. Historically, Spanish has been the most powerful catalyst for the flow of TV shows, specials, and products between the U.S. Hispanic and Latin American markets inasmuch as it helped consolidate the unity of a common Spanish-speaking community against the Anglo media world, and functioned as an attenuating barrier against the already high flow of U.S. cultural products and programming into the Hispanic market (McAnany and Wilkinson). Namely, in contrast to exports to Canada, U.S. media exports to Latin America are translated, and this provides additional room for the indigenization of these products through language.
14. See Cristina Saralegui for a description of Cristina's fight to have her Cuban accent validated in the networks and Negrón Cruz for an interview with Raymond Arrieta who describes how he finally learned to speak with "caution" on the airwaves.
15. Thus, my intellectual Latino friends and myself use Spanglish playfully and as a vibrant linguistic resource, but this use is never a sign of linguistic deficiencies. Class and linguistic ability in both English and Spanish render its use safe in ways that are not common among most Latinos.
16. These dynamics are not unique to the case at hand. Writers have long noted that language ideologies construct language to be emblematic of particular peoples and of their national distinctiveness. Based on dominant Herderian Western-based nationalist premises, language that is creolized, mixed or the product of code switching is therefore seen as embodying the decay of its speakers and their culture. Nor surprisingly, activists of minority languages end up reproducing the same essentialized politics that reproduce distinctions and hierarchies in language use in the process of resisting their own marginalization (Woolard).
17. See for instance the comments of former Telemundo president, Peter Tortorici (in Lizette Alvarez).
18. Episode 106, "Chicago Rendevous," 1998.
19. See Ellis Cashmore and Clara Rodríguez, respectively, for a discussion of the buffoon stereotype as it applies to blacks and Latinos.

Works Cited

Alvarez, Lizette. "Border Crossing: Speaking Double," *New York Times Magazine* (September 20, 1998): 46–47.

Ang, Ien. *Living Room Wars: Rethinking Media Audiences for a Postmodern World.* London: Routledge, 1996.

Appadurai, Arjun. "Disjuncture and Difference in the Global Cultural Economy," *Public Culture* 2.2 (1996): 1–24.

Avila, Alex. "Trading Punches, Spanish-Language Television Pounds the Competition in the Fight of Hispanic Advertising Dollars," *Hispanic* (January-February 1997): 39–44.

Barraza, John. Letter to the Editor. *Hispanic Magazine* (June/July 1998): 8.

Belejack, Barbara. "Aztec Time," *Latin Trade* (July 1998): 46.

Cashmore, Ellis. *The Black Culture Industry.* London: Routledge, 1997

Dávila, Arlene M. "Latinizing Culture: Art, Museums and the Politics of U.S. Multicultural Encompassment," *Cultural Anthropology* 14.2 (1999): 180–202.

Esparza, Elia. "Will Telemundo Become Assimilated?" *Hispanic* (January/February 1998): 18–22.

———. "Must Sí TV." *Hispanic* (May 1998): 20–28.

Flores, Juan (with George Yúdice). "Living Borders/Buscando America: Languages of Latino Self-Formation," *Divided Borders.* Ed. Juan Flores. Houston, TX: Arte Público Press, 1993: 199–224.

Gray, Herman. *Watching Race: Television and the Struggle for Blackness.* Minneapolis, MN: University of Minnesota Press, 1995.

Gutiérrez, Félix. "Mexico's Television Network in the United States: The Case of Spanish International Network," Proceedings of the Sixth Annual Telecommunications Policy Research Conference. Ed. Herert Dordick. Lexington, MA: Lexington Books, 1979.

Hill, Jane H. "Mock Spanish: A Site for the Indexical Reproduction of Racism in American English," Language and Culture Symposium (University of Chicago) Available at <http://www.cs.uchicago.edu/l-c/archives/subs/hill-jane/> 1999.

McAnany, Emile and Kenton T. Wilkinson. *Mass Media and Free Trade.* Austin, TX: University of Texas Press, 1996.

Mejia, Victor. "Sony Introduces Telemundo's New Management," *Hispanic Magazine* (October 1998): 16.

Miller, Toby. "Television and Citizenship: A New International Division of Cultural Labor?" *Communication, Citizenship, and Social Policy: Rethinking the Limits of the Welfare State.* Eds. Andrew Calabrese and Jean-Claude Burgelman. New York: Rowman & Littlefield Publishers, Inc., 1999: 279–292.

Naficy, Hamid. *The Making of Exile Culture.* Minneapolis, MN: University of Minnesota Press, 1993.

Negrón Cruz, Wanda. "Un Profesional Serio Entre Risas (A Non-Serious Yet Serious Professional)," *Cristina* 8.11 (1998): 56–57.

Ong, Aihwa. "Cultural Citizenship as Subject Making: Immigrants Negotiate Racial and Cultural Boundaries in the United States," *Race, Identity and Citizenship.*

Eds. Rodolfo Torres, Louis F. Mirón, and Jonathan Xavier Inda. Malden, MA: Blackwell, 1999: 262–293.

Rodríguez, America. "Objectivity and Ethnicity in the Production of the Noticiero Univision," *Critical Studies in Mass Communication.* 13 (1996): 59–81.

Rodríguez, Clara. *Latin Looks: Images of Latinas and Latinos in the U.S. Media.* Boulder, CO: Westview Press, 1997.

Rina Benmayor and William Flores, eds. *Latino Cultural Citizenship, Claiming Identity, Space and Rights.* Boston, MA: Beacon Press, 1997.

Saralegui, Cristina. *Cristina, Confidencias de Una Rubia.* (*Cristina, My Life as a Blond*) New York: Warner Books, 1998.

"Se Quejan de Cadenas Hispanas de Televisión (They Complain about Hispanic TV Stations)," *El Diario* New York City (February 3, 1998): 20.

Segal, Daniel and Richard Handler. "U.S. Multiculturalism and the Concept of Culture," *Identities: Global Studies of Culture and Power* 1.4 (1995): 391–408.

Univision. "The U.S. Hispanic Market in Brief." Univision Media Kit, 1999.

Urciuoli, Bonnie. *Exposing Prejudice: Puerto Rican Experiences of Language, Race and Class.* Boulder, CO: Westview Press. 1998.

Wilk, Richard. "Consumer Goods as Dialogue about Development: Colonial Time and Television Time in Belize," *Consumption and Identity.* Ed. Jonathan Friedman. Chur, Switzerland: Harwood Academic Publishers, 1995: 97–118.

Wilkinson, Kenton T. *Where Culture, Language and Communication Converge: The Latin American Cultural-Linguistic Television Market.* Unpublished Ph.D Dissertation, University of Texas at Austin, 1995.

Williams, Brackette. "A Class Act: Anthropology and the Race Across Ethnic Terrain," *ARA* 18 (1989): 401–44.

Wilson, Clint and Félix Gutiérrez. *Race, Multiculturalism, and the Media: From Mass to Class Communication.* London: Sage Publications, 1995.

Woolard, Kathryn A. "Introduction: Language Ideology as a Field of Inquiry," *Language Ideologies: Practice and Theory.* Eds Bambi Schieffelin, Kathryn A. Woolard, and Paul Kroskrity. New York: Oxford University Press, 1998: 3–50.

Zentella, Ana Celia. *Growing Up Bilingual.* New York: Blackwell, 1997.

Iroquois Border Crossings

Place, Politics, and the Jay Treaty

Donald A. Grinde, Jr. (Yamasee)

In recent years, border studies have emerged as a new set of theoretical approaches to the intertwined issues of nationalism, national borders, and citizenship. This work has largely emphasized the challenges that Chicana/o and Latina/o cultures of the U.S.-Mexico borderlands pose to notions of belonging to one culturally homogeneous nation, which corresponds to the outlines of the nation-state. While border studies originally emerged from analyses of a particular geographic place (the U.S. Southwest), notions of border crossings and border identities have more recently come to be identified with members of a Latina/o diaspora that transcends the United States and Latin American nation-states. Thus, border studies' grounding in a specific place has given way to a sense of belonging that is no longer tied to place of residence and to which political limitations like nation-state borders are becoming less and less meaningful.

While these ideas to a certain extent parallel Native peoples' view of national borders as arbitrary impediments to internal unity, their notions of belonging cannot be described in terms of deterritorialized citizenship. Instead, indigenous people's disregard for national borders expresses itself in their continuing attachment to a nation that is rooted in a particular place and in struggles for sovereignty from the nation-state and its political boundaries. In the case of the Haudenosaunee (Iroquois) Confederacy, which consists of the Mohawk, Oneida, Onondaga, Cayuga, Seneca, and Tuscarora Nations, its modern day sovereignty movement generally stresses simultaneously the survival of each group as well as the maintenance of the Iroquois Confederacy in the face of pressures from both the United States and Canada. These struggles for sovereignty also find powerful symbolic expression in efforts to maintain Iroquois special border crossing rights at the U.S.-Canada frontier.

In traditional accounts of the creation of the Haudenosaunee Confederacy (dated somewhere around 1300–1550 A.D.),[1] Iroquois people embedded themselves in the land and region that encompasses Lake Ontario. Like other indigenous societies, the Iroquois Confederacy evolved out of natural kinship relationships to a specific place of origin. The Mohawk, Oneida, Onondaga, Cayuga, Seneca, and Tuscarora societies recognized that they were once one group, which had over time dispersed across a wider geographic expanse and evolved into separate communities with similar but distinctive languages. The creation of the Iroquois Confederacy responded to these groups' collective desire to remain politically related.

Indigenous societies like the Iroquois stressed collaborative political structures linked to kinship and environment, and emphasized free association or disassociation. Indigenous political constructions often lacked the ideological underpinnings (of divine right monarchy, taxation, class system, and a coercive state) that the nations created in Europe typically relied upon. Hence, accounts of the formation of the Iroquois Confederacy emphasize strife amongst related peoples and struggles to overcome that tension in a peaceful and equitable manner. Moreover, stories about the origins of the Iroquois Confederacy, the representation of their meeting place in the image of the Eastern White Pine, and the term "Oneida" itself (which means "people of the standing stone") contain references to a specific place that have helped to reinforce the identity and polity of the Iroquois despite white intrusions into their region.[2]

The onslaught of European colonialism threatened not only the bonds of kinship within the Iroquois Confederacy, but also Iroquois people's relations to their environment. By the seventeenth century, French, Dutch, and English colonials had pressured Iroquois leaders to ally with them in their various colonial rivalries, which caused internal divisions among the Iroquois people and ultimately led to the creation of reservations in Canada and the United States after the American Revolution. The institution of reservations also entailed the confiscation of the remaining land base of the Haudenosaunee, which was coveted by the colonial powers. Further colonial intrusions eventually resulted in the division of the Iroquois' traditional domain through the imposition of first European colonial and then Euroamerican political and economic boundaries. The loss of land and the internal separation of the Iroquois community posed not only a myriad of economic problems but also significantly affected their culture and kinship networks.

At the same time that colonial borders were imposed upon the Iroquois, treaties between the United States and England provided for the free passage of Native peoples. The Jay Treaty of 1794 guaranteed the Iroquois the right to freely cross the U.S.-Canada border without having to pay duties on trade goods. In 1814, at the Treaty of Ghent that concluded the War of 1812, the United States and Great Britain further reinforced these rights. In this treaty, both sides agreed "to restore to such tribes or nations, respectively, all possessions, rights and privileges which they may have enjoyed or been entitled to in . . . [1811] previous to such hostilities" (1814 8 Stat. 218). Both the Jay Treaty of 1794 and the 1814 Treaty of Ghent thus firmly guaranteed the Iroquois the right of free passage across the U.S.-Canada border.

Originally, the two treaties were drawn up for rather pragmatic reasons: The English wanted the Iroquois on the U.S. side to continue selling furs in Canada so that they could purchase trade goods there. The United States was more than willing to grant and maintain these special border-crossing rights for the Iroquois. But since the assault on treaty rights, which began in the nineteenth century and continues today, the struggle for the maintenance of border crossing provisions has become one of the most vivid symbols of the Iroquois' fight for sovereignty and against the abrogation of treaties by both the U.S. and the Canadian governments. These struggles have gained further urgency since the 1994 passage of NAFTA. Expanding the long history of the two North American nation-states' assaults on treaty rights, this agreement codifies the lack of corporate interest in enabling the free passage of Native peoples across international borders.

Iroquois treaty rights began to be challenged as early as the nineteenth century in attempts to transform Native peoples on both sides of the border into either American or Canadian citizens without special border crossing privileges. In 1924, the U.S. government unilaterally passed the American Indian Citizenship Act, which conferred citizenship status onto the Haudenosaunee people without as much as asking their consent. Iroquois leaders like Chief Clinton Rickard (Tuscarora) feared that the U.S. government attempted to end Iroquois treaty status by making them into taxpaying citizens who "could sell their homelands" and by depriving them of their special border crossing rights (Graymont 53).

Similar to its U.S. counterpart, the Canadian government changed its stance towards Indians in the late nineteenth century and decided to force citizenship onto Canadian Indians. The object of this "enfranchisement" was to incorporate Native peoples into white society and to destroy First Nation communities. Compared to the more explicit acts of the U.S. government, Canada tried to lure Native people into citizenship by offering money to anyone who signed up. After they became Canadian citizens, Native people were removed from the tribal rolls and could no longer own Reserve lands. As Canadian citizens, they would also lose their special border crossing rights. A Canadian Native who chose to become a citizen was then entitled to one per capita share of his band plus 20 times the annual per capita treaty payment for that band (Graymont 173). But between 1876 and 1918, only 102 Native people chose this option, the majority of whom were married to whites (Frideres 12). The few Native people who became citizens often squandered their money and then returned to their communities as paupers.

In addition to enfranchisement, at the end of the nineteenth century the Canadian government also pushed for the institution of electoral policies, which were designed to break up the Iroquois and end their special treaty rights. At that time, a small but vocal minority of "Christian" Iroquois who felt slighted by the elders and left out of the traditional chieftainship system allied with the Canadian government to break the traditional League of the Iroquois in the name of electoral reform. In 1899, the Royal Canadian Mounted Police (RCMP) went to the Akwesasne Mohawk Reservation to force an election amongst the traditional Mohawks. Michael Mitchell, Mohawk, gives the following account of this affair:

> At 4:00 A.M. on May 1, 1899, Colonel Sherwood . . . came to Akwesasne, leading a contingent of police . . . They occupied the Council Hall, where they sent a message to the chiefs to attend a special meeting . . . As the chiefs walked into the council office, they were thrown to the floor and handcuffed. One of the women notified Head Chief, Jake Fire, and as he came through the door demanding the release of his fellow chiefs he was shot twice, the second shot being fatal. The police marched their prisoners to the tugboat and left the village. Jake Fire was shot in cold blood while fighting for Mohawk Indian government . . . The seven chiefs were imprisoned. Five of them were kept in jail for more than a year . . . Immediately after this affair, the representatives of the government took fifteen Indians over to Cornwall and provided them with alcohol. The Indian agents told them to each nominate one of the others present. This was how the elective government under the Indian Act system was implemented at Akwesasne. (Mitchell 118–119)

Mitchell's account vividly exposes the undemocratic way in which Canada introduced electoral "democracy" to the Akwesasne Mohawks.

In its combined form, the policies of "enfranchisement" and "democracy" threatened the very survival of the Iroquois as a nation as well as the maintenance of their treaty rights. By the 1920s, Deskaneh (or Levi General), a traditional Cayuga Chief at the Six Nations Reserve in Brantford, Ontario, began to oppose these policies. Like many other Iroquois traditionalists, he felt that new Canadian legislation, such as the Canadian Compulsory Enfranchisement and Soldier Settlement Acts, would result in the loss of Reserve lands and in the dissolution of the Iroquois people (Shimony 166). Duncan Scott Campbell, then Canadian head of Indian Affairs, was the architect of the new Indian policy, which was in effect from 1913 to 1932. Even though Campbell had a romantic interest in Native people, he was quite forceful with actual Indians. He went on record, stating that he wanted to "get rid of the Indian problem" (Asch 62–63) and that he felt that Native people were "ready to break out at any moment in savage dances; in wild and desperate orgies" (Titley 31–32).

In 1922, while the Canadian government was discussing sovereignty issues with the chiefs at the Six Nations Reserve, the RCMP raided the Reserve under the pretense of looking for moonshiners. After shots were exchanged, talks were cut off. Subsequently, Duncan Scott Campbell decided to station a detachment of the RCMP just a few yards away from the traditional Iroquois Council House in Oshweken. On October 7, 1924, armed police forced their way into the traditional council house at Oshweken and read a decree that dissolved the Six Nations traditional confederacy. They also seized documents that related to sovereignty issues going back to the time of Joseph Brant (Veatch 324). The RCMP had identified the traditional chiefs as "troublemakers," raided their homes, and sent the leaders they could find to jail. These leaders were to be tried in Anglo courts before juries with no Native people.

In the face of such repression, Iroquois traditionalists began to fight for the maintenance of sovereignty and treaty rights in struggles for which border crossing rights would emerge as a powerful symbol.[3] These leaders began to appeal to England, the original signer of both the Jay Treaty and the Treaty of Ghent. Levi General had fled to the United States while his supporters at home raised money to send him to Great Britain. General traveled to London in 1921 and voiced his grievances there. But the British colonial secretary argued that with the transformation of the British colonies into the Dominion of Canada in 1867, Canada had taken over the jurisdiction over England's former allies—the Six Nations. General, however, insisted that the abrogation of treaty rights by Canada nevertheless required the consent of the Six Nations Confederacy in Canada.

In 1923, General again traveled to England, and then to Scotland and Switzerland, where he made his case before the League of Nations. General and a white Rochester lawyer named George P. Decker urged the League Council to recognize small autonomous American Indian nations like the Iroquois as independent nations. At the same time that their appeal was turned down, Canada also began to argue that Iroquois treaties were mere "scraps of paper" that did not need to be enforced. Undaunted, General and Decker managed to get into print quotes from the original treaties and other documents, such as a 1912 statement by the British government asserting that "[t]he documents, Records, and Treaties between the British Governors in former times, and your wise forefathers, of which, in consequence of your request, authentic copies are now transmitted to you, all establish the Freedom and Independency of your Nations" (qtd. in Mohawk 21).

In the meantime, the Canadian government did everything in its power to divide the Iroquois internally by pushing the issue of electoral reform. The government supported a representative government referendum that was advocated by a small group of Iroquois leaders who were opposed to the traditional chiefs. Although only about ten percent of the Six Nations Reserve supported the system of elected chiefs, the motion carried because many traditional Iroquois boycotted the election. On September 17, 1923, an elected council was established; thus two rival governments began to coexist on the Six Nations Reserve. This rivalry increased tensions at the Reserve since the Canadian government forcibly supported the elected council. In 1924, the Canadian government made its support explicit by "abolishing" the traditional Haudenosaunee government at the Six Nations Reserve.[4]

On November 4, 1924 with the Canadian police about to arrest him, Levi General traveled again to the League of Nations in Switzerland to enlist its aid in resisting Canadian attacks on the sovereignty and treaty rights of the Six Nations. But once again, the League of Nations ignored his pleas (Graymont 62). Despairing, Levi General left Europe in early 1925 and went to stay with his friend and member of the Confederacy, Chief Clinton Rickard (Tuscarora), in Rochester, New York. Like other traditional Iroquois chiefs in Canada, General went on record to characterize electoral reform as attempts by the Canadian government to "forc[e] foreign laws on [the Iroquois], refusing to let them run their own affairs."[5]

General's friend, Chief Clinton Rickard, secured him a speaking engagement before the Red Jacket Lodge of Masons in Lockport, New York, on March 5, 1925. General's speech concerning sovereignty issues on the Six Nations Reserve was well received as was his radio address on the same topic on March 10, 1925 (Graymont 63). In his radio address, General asserted his distaste for Indian policies north and south of the U.S.-Canada border. He said:

> Over in Ottawa, they call that Policy "Indian Advancement." Over in Washington, they call it "Assimilation." We who would be the helpless victims say it is tyranny. If this must go on to the bitter end, we would rather that you come with your guns and poison gas and get rid of us that way. Do it openly and above board. Do away with the pretense that you have the right to subjugate us to your will. Your governments do that by enforcing your alien laws upon us.[6]

After his radio address, General fell ill with pneumonia and was taken to a hospital in Rochester (Shimony 165). Eight weeks later, after treatment by nine doctors, General requested that he be taken to Chief Clinton Rickard's home on the Tuscarora Reservation outside of Niagara Falls, New York. At

Rickard's home, two medicine men treated General for a week, and soon General's health began improving. When General's friends from the Six Nations Reserve came for a visit, the Canadian government sent the RCMP to Rickard's house as they believed that General was hatching a plot against them (Graymont 63–64).

Around the same time, the U.S. government enacted its racist Immigration Act of 1924. Section 13c provided that "No alien ineligible to citizenship shall be admitted to the United States."[7] Although this section was primarily designed to exclude immigrants from the Asian subcontinent and Europe as Claudia Sadowski-Smith discusses in more detail in her essay in this collection, the Immigration Service was anxious to also apply it to Native people entering the United States from Canada. By June 1925, General's wife, children and brother as well as his friends and medicine men in Canada were barred from visiting him in New York. General told Rickard, "I'm done," since medicine men were no longer able to cross the U.S.-Canada border to treat him. General died very soon after this observation on June 27, 1925, with the official cause of death listed as pulmonary hemorrhage.[8] General's body was taken back to the Upper Cayuga Longhouse on the Six Nations Reserve in Canada, and two thousand mourners as well as the RCMP came to his funeral. Planning to comb the burial speeches for seditious content, the RCMP found, however, that everything was said in Iroquois languages (Wright 326).

In some ways, General's experience galvanized Chief Clinton Rickard's resolve to restore border crossing rights for Iroquois people on both sides of the border. The catalyst for the renewal of struggles for Iroquois border crossing rights was the 1925 arrest of Paul Diabo, a Kahnewake Mohawk steelworker, in Philadelphia. The Immigration Service arrested Diabo as an illegal alien under the 1924 Immigration Act. But Diabo argued that he was a Native American and, as such, was protected from both United States and Canadian immigration restrictions by the 1794 Jay Treaty.

To provide legal representation for Diabo, in 1926 Rickard, Sophie Martin (Mohawk), and David Hill (Mohawk) organized the Six Nations Defense League, which was soon broadened in scope and renamed the Indian Defense League. As a result of his able defense, Diabo's arguments prevailed in federal court and set a precedent that still allows Iroquois to cross the U.S.-Canada border without restrictions today. Soon the Indian Defense League expanded its activities to more generally protest the violation of treaty rights on the border. In memory of the Diabo case, the Indian Defense League began celebrating the 1925 court victory through an annual Border Crossing Ceremony in July (Snow 193–194). Moreover, it also encouraged Iroquois to insist on their legal rights by refusing to pay customs duties and to surrender their passports at the U.S.-Canada border.[9] With regard to Iroquois on the U.S. side, these rights were outlined in Article 3 of the Jay Treaty. It declares that:

> Indians dwelling on either side of the said boundary line, [shall] freely . . . pass and repass by land or inland navigation, into the respective territories and countries of the two parties, on the continent of America . . . No duty of entry shall ever be levied by either party on peltries brought by land, or inland navigation into said territories respectively, nor shall the Indians passing and repassing with their own proper goods and effects of whatever nature, pay for same impost or duty whatever. (1791 7 Stat. 47–48)

Article 9 of the Treaty of Ghent (1814) reinforced these rights, stating at the conclusion of the War of 1812 that American Indian nations shall be restored to "all the possessions, rights and privileges which they enjoyed . . . previous to these hostilities" (1814 8 Stat. 218). Throughout the early twentieth century, many Iroquois and Indian Defense League of America members were imprisoned for asserting their legal border crossing privileges. Their determined efforts—manifested through protests and annual Jay Treaty renewal ceremonies along the U.S-Canada border—affirmed American Indian treaty rights throughout the 1920s, 1930s, and 1940s.

But the 1950s witnessed renewed assaults by both the U.S. and the Canadian governments on Iroquois border crossing rights. Mohawks at the Akwesasne Reservation (split in two by the U.S.-Canada border) were faced with conflicting national jurisdictions. While Mohawks on the Canadian side were free to cross to the United States and sell crafts at county fairs, Mohawks on the American side (on the assumption that they were American citizens) were forbidden by the Canadian government to work in Canada. In 1959, a Native person from the American side who had attempted to work in Canada was found murdered on the Canadian side (Wilson 91).

In the meantime, the Indian Defense League continued to energize traditionalist Iroquois resistance and to nurture the League of North American Indians. One of the Haudenosaunee Leaders to emerge at this time was Wallace "Mad Bear" Anderson (Tuscarora). In 1957, Anderson led Iroquois protests against payment of New York State income taxes. Several hundred Iroquois people marched on the Massena, New York, state office building and burned summonses issued for unpaid taxes. In March 1959, Anderson was present at the Six Nations Reserve in Brantford, Ontario, when a declaration of sovereignty was promulgated by Iroquois traditionalists. In response, the Canadian government ordered the RCMP to occupy the Six Nation Reserve's Council House (Johansen *Encyclopedia,* 144).

During the 1950s, the Canadian government instituted yet another attack on Iroquois treaty rights by attempting to abrogate them through federal court decisions. On May 23, 1954, the Canadian Supreme Court ruled that Canadian Native Americans must pay duty on goods brought in from the United States. The ruling asserted that the Jay Treaty was actually terminated by the War of 1812—which resulted in the Treaty of Ghent—and that the Jay Treaty thus no longer applied to Canadian Indians. This decision allowed Canadian customs officials to immediately deny Iroquois special border crossing rights.[10]

The court ruling was driven by a renewed Canadian effort to "integrate" Native people into the larger society by eradicating treaty rights and by relocating Indians to urban areas. These policies resembled federal Indian assimilation and urbanization politics instituted in the United States between 1946 and 1960. Given this political context, Canadian courts were unwilling to affirm rights particular to Native peoples (Nichols 293–294). However, the Canadian Supreme Court decision also renewed the resolve of Iroquois on both sides of the border to continue their struggle for treaty rights guaranteed to Canadian Iroquois under the Jay Treaty. In 1968, Mohawks gathered on the international bridge at Cornwall to protest the duties imposed on commodities they brought in from the United States, claiming that the Canadian government violated the original terms of the 1794 Jay Treaty ("The Indians" 10).

These legal controversies surrounding treaty rights intersected with the Canadian government's attempt at instituting policies of termination. Resembling 1950s termination attempts in the United States, Canadian policies were articulated in the Trudeau government's "White Paper" on American Indian affairs. Although ultimately rejected by Native American peoples and subsequently withdrawn by the Trudeau government, the "White Paper" called for Indian "equality" or "non-discrimination" through legislation that would terminate the Indian administration in Canada's federal government and instead transfer to individual provinces responsibilities for delivering services to Native peoples. The paper suggested the elimination of the Indian affairs bureaucracy and of the Department of Indian Affairs and Northern Development, keeping only trustee functions for tribal lands, so that, within five years, Canadian Indians were to receive the same services at the same locations as other Canadians.

At a time when the United States was admitting that its termination of U.S. Indian treaties and federal services in the 1950s was a disaster, the policy inscribed in the Trudeau "White Paper" represented a very antiquated approach. But, as Sally Weaver has stated, the "long range goal of terminating the special treatment of Indians had been a part of [Canadian] governmental policies since the 1830s" (4–5). In response to these policies, Iroquois and other First Nations in Canada created a new form of American Indian nationalism, which was "unparalleled in Canadian history [and] contribut[ed] to the founding and growth of native organizations in Canada" (Weaver 4–5). After the "White Paper" was issued, *Akwesasne Notes* (the first Iroquois-owned and Iroquois-operated news magazine) began devoting more press coverage to policies coming from Ottawa. Simultaneously, Mohawks like Ernie Benedict, who had been active in earlier border crossing activities, eventually forced the Trudeau government to withdraw its proposals for elimination of Canadian federal services to First Nations peoples ("End" 1).

Originally inspired by opposition to termination policies and, more generally, to the two North American nation-states' desires to abrogate their treaties with Native peoples, in the 1970s Iroquois resistance broadened to also develop sophisticated critiques of the emerging multinational corporate system. During this time, the Iroquois Confederacy at Onondaga began to issue their own passports for international travel and "Red Cards" for additional identification. According to Oren Lyons (Onondaga), these passports and "Red Cards" are now accepted by governments around the world for traveling Iroquois people.

In 1977, Haudenosaunee leaders issued a public statement to the world, asserting that corporate interests expand upon the colonial attack of nation-states on treaty rights. The statement argues that, above all else, both efforts have managed to marginalize Native peoples economically. It declares that,

> [a]lthough treaties may often have been bad deals for the Native nations, the United States and Canada chose not to honor those which exist because to do so would require the return of much of the economic base and sovereignty to the Haudenosaunee . . . The effect of all these policies has been the destruction of the culture and therefore the economy of the People of the Longhouse. The traditional economy has largely been replaced by the colonial economy which serves multinational corporate interests. The colonial economy is one that extracts labor and materials from the people of the Haudenosaunee for the benefit of the colonizers. The Christian religions, the school systems, the neo-colonial elective systems, all work toward these goals . . . The traditional economy is under heavy attack from many directions, and all else is an economy of exploitation. The political oppression, the social oppression, the economic oppression, all have the same face. These are tools of Genocide in North America . . . We have developed strategies to resist the economic effects of the conditions we face. But those strategies require that we revitalize our social and political institutions. (qtd. in Mohawk 106–107)

This quote shows that many Haudenosaunee leaders and intellectuals see multinationalism as just another wrinkle in the ongoing process of marginalizing Native culture, peoples, and resources. Haudenosaunee resistance to globalization pressures is multifaceted, involving not only the return to traditional economic measures but also to traditional political, cultural, and social practices. The Haudenosaunee have previously resisted changes in the face of colonial exploitation, and there is every reason to believe that they will also adapt and resist the new multinational pressures. Certainly, they have the experience, resolve, and perseverance to endure and yet maintain their distinctive way of life in spite of the outside attacks upon it.

Attacks on the traditional Iroquois economy and on their treaty rights have continued into the 1990s with the establishment of casino gambling on the Mohawk and Oneida reservations and the passage of NAFTA in 1994. Although casino gambling appears unrelated to the issue of border crossing rights, the contentious nature of gambling has split Iroquois communities internally and made some Euroamericans (as well as their federal, state, and local governmental units) less sympathetic to the necessity for the continuation of Native peoples' rights. In addition, NAFTA, which is meant to intensify the economic and political powers of multinational corporations, has further aggravated the Iroquois' enduring dilemmas concerning treaty rights, economic self-sufficiency, and border crossing rights. The old problems involving treaty rights and border crossings have become merged with the new economic order as multinational corporations have, in general, exhibited little interest in the rights of Native peoples to freely cross national borders.[11] For example, the militarization of the U.S.-Mexico border in the context of NAFTA has already made it harder for Native people of the U.S. Southwest to freely cross that boundary as their reservations are increasingly policed by the U.S. Border Patrol.

Although NAFTA and casino gambling have complicated controversies concerning place and nationhood for the Haudenosaunee, contemporary Iroquois peoples' view of the U.S.-Canada border crossing issue is still fairly uniform. One Iroquois elder summarizes this view by asserting that "the white folks . . . come along and made that split through our own land where we're already livin'. That's an artificial border, . . . and we don't even recognize it" (Boyd 205). Another elder states that every national border is artificial and that he believes that the notion of borders "is an unnatural notion and don't have no recognition in the natural world." He continued, "[t]he animals and birds, and the wind and the water, they don't pay no attention to these so-called borders" (Boyd 205). Iroquois people from a broad cultural and political spectrum still join in the annual summer border crossing ceremonies.

Essentially, Iroquois people on both sides of the U.S.-Canada border have fought to keep their homeland (the lands surrounding Lake Ontario) unified, despite the imposition of an artificial national border and despite a long history of attempts by both the United States and Canada to terminate their treaty rights. In many ways, the border crossing demonstrations in the 1920s conducted by Clinton Rickard and the Indian Defense League can be seen as direct precursors of the twentieth-century American Indian protest movement. To draw attention to the Diabo Case which was in U.S. federal court at the time, Rickard and other Iroquois leaders organized a meeting at the

Peace Bridge in Buffalo on August 6, 1927, to dramatize controversies surrounding issues of border crossing (Graymont 83–85). Iroquois' refusal to pay custom duties or to surrender their passports while crossing the U.S.-Canada border dramatized governmental noncompliance with the Jay Treaty and the Treaty of Ghent. Such actions by Iroquois leaders and activists in the early twentieth century paved the way for other Native nations to reassert their treaty rights in the latter half of the twentieth century.

In many ways, these demonstrations have also fostered a renewal of Iroquois sovereignty rights and of Iroquois traditional culture. Having inhabited the lands around Lake Ontario for thousands of years, Iroquois see the U.S.-Canada border as impeding their unity. For them, the border serves foremost as a symbol of their oppression by colonizing national governments that have sought to destroy and/or to ignore their existence. The border for the Iroquois people is an "unnatural notion" and will continue to be a point of protest and contention as long as it hinders the free passage of the Iroquois through their traditional homeland.

The border is a clear manifestation of colonialism and of the continuing domination of Native peoples. After England abandoned its colonial designs in North America, the United States and Canada have continued colonial Native American policies long after gaining their respective independence from England. The border impairs the unity of the Iroquois people and negatively affects some of the most basic aspects of their economic, community, and political lives. NAFTA constitutes just another complication in a long line of colonial transgressions by nation-states as today's multinational corporations are pursuing their own economic interests with little sensitivity to the lives of Iroquois people around Lake Ontario. The way that the United States and Canada define the lands and boundaries around Lake Ontario violates the Iroquois understanding of their own sense of place. These differing perceptions have produced economic, political, and social conflicts for a long time, and there is every reason to believe that those unresolved differences will persist as long as Canada and the United States ignore the treaty rights and the sovereignty of the Iroquois people on both sides of the border.

Notes

1. There is great debate about the exact beginnings of the Iroquois Confederacy (from 1300–1550 A.D.). Iroquois people today tend to date it in the 1300s while Euro American scholars say late 1400s or early 1500s. Neither side has much data to support their claims although the stories of the founding of the Iroquois Confederacy uniformly assert that it was at the time of a solar eclipse and the only one is about 1350 A.D. for that era and area. I tend to think that the late fourteenth century is a good estimate.

2. See Tehanetorens (Ray Fadden) for an account of the traditional imagery of the Iroquois, 1–14.
3. See Deskaneh (Levi General) 10–17 and Noon 59–65.
4. See Graymont 61–62 and Mohawk 103.
5. See Graymont 58 and Mohawk 18.
6. Quoted in Mohawk 27, "Deskaheh," and in Rostkowski.
7. See *U.S. Statutes at Large* 1925 43:159 and Graymont 65.
8. Illegal Chinese immigrants were also posing as Native Americans to gain entry into the United States during the era of Chinese Exclusion, see Erika Lee, Graymont 65, Mohawk 32, and Shimony 165.
9. See Johansen *Encyclopedia* 143 and Graymont 76–77.
10. See Johansen *Encyclopedia* 143–144 and "Indian Chiefs" 13.
11. For a more in-depth discussion of the relationship between corporations and treaty rights, see Nichols 320–324, Trafzer 440–443, Johansen *Encyclopedia* 145–147, Johansen *Life* 28–29, and Grinde 171–201.

Works Cited

Asch, Michael. *Home and Native Land.* Toronto: Methuen Press, 1984.

Boyd, Doug. *Mad Bear.* New York: Simon & Schuster, 1994.

Deskaneh (Levi General) and the Six Nations Council. *The Redman's Appeal for Justice.* Brantford, Ontario: Wilson Moore, 1924.

"Deskaheh: Iroquois Statesman and Patriot," *Six Nations Indian Museum Series.* Akwesasne, NY: Akwesasne Notes, n.d.

"End Special Status of Indians, Canada's Decision," *Akwesasne Notes* 1 (June 1969): 1.

Frideres, James S. *Native Peoples in Canada.* Toronto: Prentice Hall, 1988.

Graymont, Barbara, ed. *Fighting Tuscarora: The Autobiography of Chief Clinton Rickard.* Syracuse, NY: Syracuse University Press, 1973.

Grinde, Donald A. Jr. and Bruce E. Johansen. *Ecocide of Native America: Environmental Destruction of Indian Lands and Peoples.* Santa Fe, NM: Clear Light Publishers, 1995.

"Indian Chiefs Fight Canadian Duty on Goods," *Chicago Tribune* (September 24, 1954): 13.

"The Indians Rebel Again," *Chicago Daily News* (December 30, 1968): 10.

Johansen, Bruce E., ed. *The Encyclopedia of Native American Legal Tradition.* Westport, CT: Greenwood Press, 1998.

Johansen, Bruce E. *Life And Death in Mohawk Country.* Golden, CO: North American Press, 1993.

Lee, Erika. "Crossing Borders and Race: Illegal Chinese Immigration During the Exclusion Era," Presentation at the 1999 American Studies Association Meeting, Montreal, Canada.

Mitchell, Chief Michael. "Akwesasne: An Unbroken Assertion of Our Sovereignty," *Drum Beat.* Ed. Boyce Richardson. Toronto: Assembly of First Nations, 1989.

Mohawk, John. ed., *Basic Call to Consciousness* Summertown, TN: Book Publishing Company, 1991.

Nichols, Roger L. *Indians in the United States and Canada: A Comparative History.* Lincoln, NE: University of Nebraska Press, 1998.

Non, John A. *Law and Government of the Grand River Iroquois.* New York: Viking Fund Publications in Anthropology, 12, 1949.

Rostkowski, Joelle. "The Redman's Appeal for Justice: Deskaheh and the League of Nations," *Indians and Europeans.* Ed. Christian F. Feest. Aachen, Germany: Edition Herodot, 1987.

Shimony, Annemarie. "Alexander General, 'Deskahe,' Cayuga-Oneida, 1889–1965," *American Indian Intellectuals* Ed. Margot Liberty. St. Paul, MN: West Publishing Co., 1978.

Snow, Dean. *The Iroquois.* Cambridge, MA: Blackwell Publishers, 1994.

Tehanetorens (Ray Fadden). *Tales of the Iroquois.* Rooseveltown, NY: Akwesasne Notes, 1976.

Titley, E. Brian. *A Narrow Vision.* Vancouver: University of British Columbia Press, 1986.

Trafzer, Clifford E. *As Long as the Grass Shall Grow and Rivers Flow: A History of Native Americans.* New York: Harcourt Publishers, 2000.

Veatch, Richard. *Canada and the League of Nations.* Toronto: University of Toronto Press, 1975.

Weaver, Sally. *Making Indian Policy: The Hidden Agenda, 1968–1970.* Toronto: University of Toronto Press, 1981.

Wilson, Edmund. *Apologies to the Iroquois.* New York: Vintage Books, 1959.

Wright, Richard. *Stolen Continents: The Indian Story.* London: Random House, 1992.

3. Border Alliances

Las Voces de Esperanza/ Voices of Hope

La Mujer Obrera, Transnationalism, and NAFTA-Displaced Women Workers in the U.S.-Mexico Borderlands [1]

Sharon A. Navarro

On June 11, 1997, a group of Mexican American women blocked the Zaragoza International Port of Entry—one of El Paso's busiest commercial bridges with Cuidad Juárez, Mexico. By stretching a rope across the border crossing's street entrance, the women successfully stopped traffic for more than an hour. The protestors were organized by La Mujer Obrera (the Working Woman) (LMO), an independent organization of Mexican immigrant women workers from the El Paso garment industry. Prior to the passage of NAFTA, jobs in El Paso's manufacturing had been largely held by Mexican American women, many of whom originally migrated from Mexico in search of a better life. As a result of NAFTA's trading and commerce policies, however, El Paso factories began moving across the border, and the Mexican American female labor force was replaced by Mexican women *maquiladora* workers. The women hoped that their action on the border would bring the plight of NAFTA-displaced workers to the attention of Secretary of Treasury Robert Rubin, then the head of the federal department responsible for a variety of NAFTA programs. But instead of being able to meet with him, his representative, or other state or local authorities, protesters were arrested and charged with obstructing highway commerce.

The Zaragoza protest made the LMO and its struggles visible so that a January 1999 *El Paso Times* article by Mike Mrkvicka listed two of the organization's leaders—Cindy Arnold (development coordinator for *El Puente* Community Development Corporation) and Guillermo Glenn (coordinator for the *Asociación de Trabajadores Fronterizos*) as among the top ten most influential leaders shaping El Paso's economy. LMO has become the most well-known and well-established Mexican American community-based organization in El Paso. As a Mexican American women's worker advocacy organization it is rooted in Mexican culture. In the era of globalization with its growing emphasis on hemispheric integration and disappearing nation-state boundaries, LMO's closure of a major international commercial bridge exhibited what I call a form of "oppositional nationalism." The LMO's actions created a transnationally shared cultural space by employing central icons of Mexican culture to mobilize its constituents in appeals to U.S. state and government institutions. At the same time, the organization also consciously opted *not to* establish cross-border coalitions with labor organizations in Mexican *maquiladoras.* For LMO, a transnationally shared Mexican ideology and culture has become the basis of and justification for limiting actions to the borders of the U.S. nation-state, that is, for the struggle to have issues of previously excluded groups included into local and national U.S. politics.

This essay takes an interdisciplinary approach to explain LMO's complex transnational relationship with Mexico within the larger context of globalization as manifested in NAFTA. Drawing on literature from anthropology, history, sociology, feminist studies, and political science as well as from a wealth of new data obtained through my fieldwork,[2] the essay explores the contradictory role LMO plays by embracing Mexican national culture, on the one hand, while at the same time foregoing the opportunity to form cross-border coalitions with labor organizations in Mexico. LMO's reinterpretation of Mexican culture north of the U.S.-Mexico border is suggestive of a growth in *cultural cross-border integration,* while it also exhibits increased *political and economic separation* embodied in the strengthening of national borders.

Mexican Women Workers in El Paso: A Gendered Look at Globalization

For tens of thousands of displaced workers, the ongoing deep economic restructuring marked by NAFTA has not ushered in the new dawn of prosperity hailed by North American political leaders. The explosive growth of low paying export manufacturing jobs in the *maquiladora* sector in Mexico has been offset by an immense loss of jobs in the domestic manufacturing sector in the U.S. and Canada. Because my essay deals with a grassroots women's worker advocacy organization in the U.S.-Mexico borderlands, it does not focus on Canada.

The globalization of capital that is necessary for economic restructuring is being deliberately hastened by national governments, international institutions like the IMF and the World Bank, and by global corporations themselves. While international trade is nothing new, our system of nation-based economies is rapidly changing toward a "New World Economy."[3] At the center of this change lies a sharp increase in capital mobility. Computer, communication, and transportation technologies continue to shorten geographical distances, making possible the coordination of production and commerce on a global scale. Lower tariffs have reduced national frontiers as barriers to commerce, thus facilitating transnational production and distribution. Corporations are becoming global not only to reduce production costs, but also to expand markets, elude taxes, acquire resources, and to protect themselves against currency fluctuations and other risks (Brecher and Costello 37–80). These "other risks" are organizations like LMO that mobilize workers and demand that their interests and voices be heard in the "New World Economy."

Three hundred companies now own an estimated one-quarter of the productive assets of the world (Barnet and Cavanagh 15). Of the top 100 economies in the world, 47 are corporations, each with more wealth than 130 countries (Harison 51). International trade and financial institutions like the IMF, the World Bank, the European Union (EU), and the new World Trade Organization (WTO) have taken over powers formerly reserved for nation-states. Conversely, national governments have become less and less able to control their own economies. This new system, which is controlled by the Corporate Agenda, is not based on the consent of the governed. It has no institutional mechanism to hold it accountable to those affected by its decisions.

In general, the effects of capital mobility have been malignant on workers. An unregulated global economy forces workers, communities, and countries to compete with each other in an effort to attract corporate investment. Each body tries to reduce labor, social, and environmental costs below the others.[4] Prior to 1970, national economic policies supported regulation that set minimum standards and stimulated economic growth in an effort to counter the downward-spiraling of market economies. Since the 1970s, however, large corporations have shifted their positions from supporting such regulations to viewing them as obstacles to their emerging economic strategies. The corporations, their think tanks, and economists have begun to develop a new public policy agenda—the "Corporate Agenda"—which is designed to eliminate set minimum standards.

In the debates surrounding NAFTA, globalization is often interpreted as a homogenizing vehicle, without taking into account that women, and in particular women of color, are paying a disproportionate share of the costs of the processes of neo-liberalism. Globalization hits especially hard at racial/ethnic minorities in the U.S. and at women workers concentrated in manufacturing (Larundee 123–163). More specifically, it has profoundly changed the lives of women in El Paso, creating inequalities that interact with pre-existing class, ethnic, gender, and regional cleavages (Gabriel and MacDonald 535–563).

Job losses have been especially substantial in the apparel sector in El Paso, Texas, and in other small communities that are heavily dependent on factories since these have shut down and moved to Mexico. For a very long time, the community of El Paso was advertised as a low-wage, low-skill manufacturing paradise. A 1997 study conducted by the Greater El Paso Chamber of Commerce Foundation, Inc., characterized the city as "the low wage paying capital of the world" (El Paso Greater Chamber of Commerce 10). Roberto Franco, Director of the City's Economic Development, similarly points out that El Paso was "the jeans and slacks capital of the United States" (qtd. in Kolence 4). Most of the workers in El Paso's garment and in other manufacturing industries were Mexican American women whose subordinate social status limited their employment opportunities. As a result of NAFTA, however, thousands of Mexican American workers, many of whom had originally migrated from Mexico to El Paso, lost their jobs. These women's job skills cannot easily be transferred from the low-tech to the mid-to-high tech manufacturing jobs now entering El Paso in record numbers (El Puente CDC 6). Ninety-seven percent of the displaced workers in El Paso are Mexican American and eighty percent are women (Gilot 10B). One-third of these women head single households. Half are between the ages of 30 and 45, while the majority of the other half are older than 45. In addition, most of the affected workers are sustaining up to three other generations, including their parents, children, and grandchildren (La Mujer Obrera *Stop* 1). Using an average figure of four persons per household, the population affected by NAFTA in El Paso is estimated at 40,000. When taking only into account the number of so-called certified NAFTA- displaced workers,[5] the at-risk population is at least 80,000—13 percent of El Paso's population.

The testimonies of displaced Mexican American women workers reveal their immense financial struggles. Typically over 45 years of age, with less than a fourth-grade education in Mexico, these women have lost health insurance and other benefits because the businesses they worked for shut down production, and they are typically two to three months pass due on rent and utilities.[6] The lead organizer of the ATF, Guillermo Glenn, describes the burden globalization has foisted upon Mexican American displaced woman worker as follows:

> In El Paso, women suffer what is called a "*doble jornada*" (a double day's work) because they have two jobs, one at home and one at the factory (or school

> since there are very few factories open). They suffer all the discrimination: language discrimination, discrimination because they are Mexican, and discrimination because they are women . . . there is still a lot of discrimination in terms of their role, in terms of their decision-making, in terms of how they are treated.[7]

LMO has been working for nearly five years to make the conditions of Spanish-speaking, NAFTA-displaced women workers visible locally, regionally, and nationally. The seeds of LMO were planted in 1972, when a campaign to unionize Farah plant workers in El Paso intensified.[8] LMO was formed because women felt that the Amalgamated Clothing and Textile Workers of America, or for that matter, any union (such as the AFL-CIO), did nothing to address their needs, ignored their rights as *women* workers, and did not respect their membership as women members in the union.[9] This devaluation of women's work is deeply rooted in the history of U.S. labor unions. According to Guillermo Glenn, the old labor movement of the 1970s—including the one that organized Farah workers—"characterized itself as . . . [a] very macho kind of an organization. We (LMO) feel that the labor unions have not gotten away from that"[10] For example, male organizers in both the Amalgamated Clothing Workers of America and the AFL-CIO ignored women worker's concerns, such as sexual harassment, health care, childcare, domestic violence, political education, and verbal abuse by their employers.[11] Moreover, employers often view the income of the female workers as "extra" or "supplemental" to that of their husband's income.

Since its inception in 1981, LMO has defined itself, first and foremost, as a *women's* organization (La Mujer Obrera *Stop,* 3). The organization employs a total of six women as full-time organizers. The board of directors is made up of two men and five women, all of whom have been displaced by NAFTA. The support staff consists of a paid secretary and a grant writer as well as various interns and volunteers.

LMO combines community organizing, education, leadership development, and advocacy into a comprehensive struggle for a better quality of life for women and their families. The organization is interested in improving its members' access to decent, stable employment, housing, education, nutrition, healthcare, peace, and political liberty. Like many other women's organizations, LMO has struggled to integrate previously excluded issues into politics by pushing what are traditionally perceived to be women's concerns like childcare and healthcare (Peterson 183–206). In general, women's movements have been as much about creating women's networks as they have been about creating women's cultural spaces and empowering women, both individually and collectively (Peterson 185). In her study of Mexican American women in East Los Angeles, Mary Pardo examines how these women transform "traditional" networks and resources based on family, religion, and culture into political assets in order to defend the quality of urban life. Here, the women's activism arises out of seemingly "traditional" roles, addresses wider social and political issues, and capitalizes on formal associations sanctioned by their community. Often, women speak of their communities and their activism as extensions of their family and household responsibilities. LMO similarly serves foremost as a safe place where friendships develop, experiences are shared, and where women are educated not only about their rights as workers, but, perhaps most importantly, also about their rights as women, mothers, and spouses. To this end, LMO has established a childcare center where displaced workers may leave their children while they attend school, find employment, or go to work.[12]

As a representative of NAFTA-displaced Mexican American women workers, LMO has directed its efforts to create genuine economic alternatives primarily at city, state, and federal institutions in the United States. Recently, La Mujer Obrera has reorganized and broadened its structure (La Mujer Obrera *Progress,* 4), enabling the emergence of two new quasi-independent organizations under its corporate umbrella. The *Asociación de Trabajadores Fronterizos* (Association of Borderland Workers) has taken on worker and community organizing, direct action, and mass mobilization components of LMO's work (La Mujer Obrera *Progress,* 5). The *Asociación de Trabajadores Fronterizos* (ATF), which was launched during the summer of 1996 with the support of LMO, has expanded LMO's traditional base of women garment workers and also incorporated displaced men, workers from electronics, plastics, medical supply manufacturing, and other labor-intensive industries in El Paso. Growing from an initial membership of 20 workers to a current roster of more than 700 active members, the *Asociación* has built a countywide network of seven factories and school committees (La Mujer Obrera *Progress,* 6 & 7).

Through these various networks, more than 2,000 workers, the majority of whom are women, and a governing board of 14 displaced workers participate in LMO (La Mujer Obrera *Progress,* 26). LMO has also begun to develop and operate community economic development programs on behalf of, and with, displaced workers. In December 1997, LMO established *El Puente* Community Development Corporation as a vehicle to develop training programs, provide access to jobs, self-employment, housing, and credit as well as to plan neighborhood revitalization strategies as a means to create income and economic self-sufficiency for the workers and the organization. Moving into this arena has required that LMO negotiate and cautiously collaborate with local, state, and federal government agencies (La Mujer Obrera *Progress,* 26 & 27).

By creating a space in the political and economic arena of El Paso, LMO has already produced tangible results for displaced workers. The organization succeeded in getting a $45 million grant—the largest grant ever given by the department of labor—to institute alternatives to existing federal and state retraining programs. Existing programs entitle displaced workers to benefits while attending retraining programs within an 18-month period. Many of these programs, however, are plagued with a variety of problems, such as inadequate facilities, outdated curricula, incompetent and poorly trained teachers, culturally-insensitive administrators, and, at times, loss of benefits for up to six months.[13]

In the summer of 1999, Margarita Calderon, a researcher at Johns Hopkins University, tested 60 displaced workers at twelve different schools to see what they had learned. The results were shocking. Only two of the sixty students, some of whom had spent up to five years in various federally subsided training programs, had learned enough to handle an interview in English. Calderon, an expert in bilingual education, found that retraining schools set up for displaced workers had been using rigid teaching techniques long abandoned by better language schools, such as lectures and translation exercises (qtd. in Templin B1 & B4). In addition, there is a perception among case managers, state and local agencies, school administrators, and teachers that "the displaced workers are just there to get paid and don't want to learn."[14] Frustrated with many of the inept retraining schools, on February 1, 1999, LMO opened its own school for displaced workers. The organization hired a full-time teacher who will be teaching English, the general equivalency examination (GED), and micro-enterprise training. LMO has also established an incubator facility that will support the self-employment initiatives of displaced workers. Further, LMO is currently in the process of developing the first-ever bilingual adult education curriculum specifically designed for displaced workers in El Paso. This curriculum will be the standard for every school in El Paso that accepts displaced workers. LMO has also played an instrumental role in helping El Paso win the Empowerment Zone designation. As a result, the poorest area of the city—South Central—will receive much needed federal money for revitalization. LMO will play a role in deciding how that money is going to be utilized.

"Oppositional Nationalism": The Uses of Mexican Culture for Women's Grassroots Mobilization

LMO's mobilization of its women members to appeal to U.S. city, state, and federal institutions has been largely grounded in the creative use of elements of Mexican culture and attempts to build transnational cultural bridges across the border. In her study of two Chicano struggles in the Southwest, Laura Pulido points out that Mexican culture has played a key role in mobilization efforts both by providing familiar and meaningful guideposts and by facilitating a strengthened collective identity. The use of symbols, customs, and traditions remind people of their shared traditions and map an outline of possible achievements through collective struggle (Pulido 31–60).

Generally speaking, in the 1960s and 1970s Chicana/os re-appropriated and turned Mexican culture into symbols of resistance. In the course of building *El Movimiento,* for example, farm workers and activists publicly displayed statues and posters of *La Virgen de Guadalupe,* not only as a source of solace and inspiration, but also (and far more consciously) as an expression of pride in Mexican culture and as a tool for mobilization. By openly engaging in ritual prayer, they asserted a collective identity that stood in opposition to the larger society.[15]

Drawing on this tradition, LMO similarly utilizes *La Virgen de Guadalupe* as an expression of inspiration and as a vehicle of mobilization and consciousness raising for its members. Their statue of the Virgin was donated by a local ceramic factory that had closed down. To the women of LMO, the *Virgen de Guadalupe* thus not only represents strength, hope, and respect, but also displacement. Moreover, "she is also viewed as a woman and a mother who has suffered with the death of her son and, like her, we [as displaced workers] are suffering for our families and our children."[16]

LMO also identifies with Mexican culture through its landscaping. Outside the physical structure, the organization has planted corn (maize) stalks in place of an assortment of bushes or flowers. As Refugio Arrieta explains, "the maize is symbolic of our Mexican culture. It is a symbol of the resistance too . . . it is more like the resistance that leads to maintaining the culture here and helping it live and grow here in the U.S. and not simply being forced to abandon the Mexican culture."[17]

Moreover, LMO tries to build a transnational culture of mobilization by incorporating Mexican history into their creation of a Mexican American women's movement. Hanging from the walls of the LMO building are portraits and pictures of Emilio Zapata and Zapatista *women* rebels in symbolic remembrance of Emiliano Zapata and the more recent struggles of the EZLN (the Zapatista National Liberation Army). For the LMO:

> Zapata and the Zapatista rebels represent the struggle of resistance against neo-liberalism, specifically NAFTA. They represent in Mexico the voice of people who have been eliminated in Mexico by NAFTA. We are kindred spirits in that manner with the immigrant workers here and what the U.S. economy is doing to them and what the Mexican economy is doing to the indigenous people of Mexico.[18]

Like the Zapatistas in southern Mexico, LMO believes that NAFTA will eliminate them from an economy that was built on their backs.

LMO further uses the Mexican *corrido*[19] (ballad) as a mobilizing mechanism that reflects their political situation. For example, during Guillermo Glenn's trial over his participation in the blocking of the Zaragoza International Port of Entry, LMO members gathered outside the courthouse in protest. On the first of a three-day trial, on March 17, 1999, LMO members began to sing the "*Corrido De Los Desplazados*" (The Ballad of the Displaced), which they had composed for the occasion:

Ano de '94	In the Year of '94
Comenzo la pesadilla	The nightmare started
Se robaron los trabajos	They stole the jobs
Nos dejaron en la orilla	They left us at the edge
Los obreros en El Paso	The workers in El Paso
Recuerdan bien ese dia . . .	Remember well that day . . .

The "*Corrido de Los Desplazados*" is intended to unite and mobilize displaced workers and to raise their level of political consciousness. This type of cultural practice adds to more conventional forms of political participation, such as voting. Cindy Arnold, Coordinator of *El Puente* at LMO, summed up the organization's broad philosophy of political participation as follows: "for us, political involvement is not based on an electoral or party process, but [in] on-going dialogue with political leaders at all the different levels."[20] As a form contributing to this building of an "oppositional nationalism," the use of corridos embodies demands for justice toward NAFTA-displaced Mexican American workers and also creates a transnationally shared culture with Mexico.

But the protest actions in support of Guillermo Glenn also illustrated some of the major obstacles to LMO's organizing efforts, obstacles that are related to the roles typically ascribed to women in Mexican culture. As the local news media filmed LMO protesters, some women—at least those who were married—hid their faces behind the protest signs because their husbands would be angry with them for appearing on television. One woman said: "My husband would be humiliated by his friend if they saw me on television."[21]

In other instances, LMO activists were accused by their spouses of neglecting their family obligations after they became involved with the organization (Olvera 1990). Women in leadership roles often reveal that they have strained relationships with their husbands because they refuse to adhere to the role of a traditional Mexican wife.[22] Other women report that they are allowed to participate in LMO as long as they also perform the duties expected of them as mothers, housekeepers, and wives.[23] Some women mention having difficulties getting away from housework and their husbands. Maria Acosta, for example, describes the problem she has with her jealous husband as follows: "Every time I get ready to come to the meetings, my husband gives me a hard time. He says that I come to (La) Mujer Obrera because I am either looking for a boyfriend or I come to meet my boyfriend."[24] According to Acosta's husband, there was no other reason for her to attend meetings at the LMO, despite the fact that they were helping her to get into a retraining program.

Building Political Cross-Border Alliances

By increasing communication and travel, processes of economic restructuring have not only facilitated the diffusion of values, knowledge, and ideas, but have also enhanced the ability of like-minded groups to organize across national boundaries. The strength of such groups rests on their ability to articulate a powerful set of shared values, to harness a growing sense of international consciousness, and to respond to failures that occur locally and globally (Hurrell and Woods 447–470). For example, like others around the globe, U.S. women have heard about *"Las Madres de Plaza in Argentina,"* a movement of politically inexperienced women in defense of their "disappearing" family and friends in Argentina. The women's persistent, silent, nonviolent street demonstrations and hunger strikes were essential in bringing down the repressive government as well as in gaining international support for their struggle (Navarro 241–258). There are many other examples throughout Latin America where women have mobilized around a number of issues and educated themselves by way of the increased globalization of communication.

By incorporating elements of Mexican history and culture into their strategies of political mobilization directed at U.S. local, federal, and state institutions, LMO's members have drawn attention to the ways in which NAFTA has placed a disastrous economic burden on them. It would appear that LMO's usage of a transnationally shared Mexican culture could also constitute the basis for an expansion of women's mobilization against NAFTA across U.S. borders, that is, within the context of transnational feminism as a political movement. After all, it is Mexican *women,* in particular, who now work in U.S.-owned factories that have often moved from U.S. border cities like El Paso into the Mexican borderlands. In many of these *maquiladoras,* women are trying to organize against the abysmal working conditions supported by the provisions of NAFTA.

But the building of such transborder political movements also requires that activists confront and recognize the tensions and differences between their various groups of women. Vasuki Nesiah, for example, calls for a "feminist internationality," "a transnational political alliance of women whose differences are acknowledged . . . and are confronted rather than ignored"

(190). The viability of future attempts to expand existing transnational coalitions will depend largely on organizers' ability to identify areas of difference and commonality. There exist important areas of shared concern for women garment workers in the United States and in Mexico since El Paso's garment industry can be seen as a forerunner of the *maquiladoras* in the Mexican borderlands. But the political, economic and social contexts in which garment workers today toil differ in many ways.

LMO is very aware of events in Ciudad Juárez and in the rest of Mexico, especially because some of its members reside in Juárez. Even though LMO knows that there exist some independent labor organizations in various Mexican cities along the border, the organizers have made a conscious decision to limit LMO's actions nationally and to make the city, local, and federal organizations address the needs of the displaced workers. LMO's organizer Cindy Arnold says that it "is challenging enough to build a movement on this side (El Paso) of the border."[25]

In addition, LMO would face serious risks if they decided to forge cross-border alliances with women *maquiladora* worker groups in Ciudad Juárez. Lead organizer Maria Flores characterizes the situation in Ciudad Juárez thus: "the way it is now, with increasing repercussions . . . it is best that we not be in the middle . . . for the protection of our organization."[26] Prior to Vicente Fox's ascendancy to the presidency in Mexico's 2000 elections, the Institutional Revolutionary Party (PRI) systemically attacked all efforts by workers to organize independent or democratic unions outside of state control. Government hostility was and still is the major reason why independent unions are such a small part of the Mexican labor movement. Mexican labor law requires that unions obtain "juridical personality" (a charter granting legal recognition) from the government. The government, in turn, has enormous power to intervene in labor movements, to arbitrarily dismiss union leaders, to declare strikes "illegal," to seize work places, to grant or withhold legal recognition, and to delay the proceedings by which workers can change union representation.[27]

With the 2000 election of a president from an oppositional party—the National Action Party (PAN)—for the first time in seventy years, labor advocates from Mexico and the United States hoped that the working conditions for the workers in the *maquiladoras* would improve. After all, the election of Vicente Fox Quezada to the presidency was supposed to mark the end of government-sponsored corruption in Mexico. During his campaign for presidency, various independent labor advocacy organizations and *maquiladora* workers had lobbied for the support of Vicente Fox. As he campaigned through border cities, he made the following promise: "My interest is for the *maquila* workers to have a better standard of living and more benefits" (Coalition for Justice).

After the presidential election, independent labor organizations and *maquiladora* workers, however, soon learned that their situation was not changing regardless of the power switch. Within weeks of Fox's ascendancy to the presidency, Francisco Cabeza de Vaca (a representative of PAN), visited various *maquiladora* workers who sought legal recognition of their independent union along the U.S.-Mexican border. He summarized the newly elected president's position by warning them that:

> You are taking the wrong way, and you should change the focus of your struggle. If you insist on forming an independent union you will never achieve anything, because you are fighting against the system . . . The union, the company and the government are involved and you will never get anything if you keep on struggling for the right of having an independent union. You should focus on issues of gender, you should form a commission of women, single mothers, asking for the reinstatement of your jobs, and I will help you to present it to the Women Commission in the Congress. Otherwise you are wasting your time. (qtd. in Coalition for Justice).

One of the most vocal labor advocacy organizations of *maquiladora* workers—the Coalition for Justice in the *Maquiladoras*—characterized the position of the newly elected president best by saying, "These actions (words) are proof of the continuing complicity between the corporations and government. They send a clear signal that nothing has changed. What happened to President Fox's campaign promises of reform?" (Coalition for Justice).

This contemporary situation has its roots in the history of Mexican labor organizations. The Confederation of Mexican Workers (CTM) was founded in the 1930s. With the support of President Lazaro Cárdenas, it quickly became the dominant labor federation in Mexico. To limit the power of the CTM and strengthen government control over the unions, Cárdenas oversaw the creation of a mixed organization of farm workers and small property owners called the National Confederation of Cooperatives (CNC) and an organization of public sector workers, the Federation of Unions of Workers at the Service of the State (FSTSE) (Alexander and Gilmore 44–49).

President Cárdenas made the CTM, CNC, and FSTSE official affiliates of the Mexican ruling party, the PRI. As a result, most union members were, until very recently, required to join the PRI. In many cases, dues were automatically deducted from workers' pay for both the union and the party. Unions of this type have become known as "official" unions because of their direct relationship with the PRI (La Botz 2–12). The PRI-CTM links are especially tight. Many CTM leaders are important PRI politicians, often controlling the PRI machine at the local level. These politicians/union bureaucrats control vast amounts of patronage and have the police at their disposal. The official unions lack democratic procedures such as membership meetings and secret ballot elections.[28]

In addition to the official unions, there are the *sindicatos blancos* (white unions)—company unions that are not independent in any real sense. These *sindicatos blancos* engage in questionable practices by creating "protection contracts," whereby official unions sell companies a contract that remains in a drawer until a real union appears on the scene. Consequently, workers may go for years without any knowledge that they are "represented," with no knowledge of their officers, no meetings, and no actual representation.[29]

Globalization, as manifested in NAFTA, may have globalized communication and travel and has made it easier to diffuse ideas, values, and knowledge. Simultaneously, however, it has made transnational alliance-building harder. LMO may shift its U.S.-centered activism and establish transnational *political* ties if the situation in Mexico significantly changes. After all, as an organization of Mexican American women, the LMO has both the cultural capital and the cultural connections that would empower it to build a transnational network of Mexican (American) women factory workers.

Conclusion

LMO's activism in response to global economic restructuring relies on a contradictory yet successful mobilization strategy. On the one hand, LMO became visible when its members literally re-enforced the U.S.-Mexico border to highlight NAFTA's disastrous effects on Mexican American women workers. On the other hand, LMO has simultaneously worked at de-stabilizing the U.S.-Mexico border culturally while choosing not to forge political cross-border alliances with labor groups in Mexico. Displaced by NAFTA, the lives of the LMO women have been disrupted and forever changed by the corporate agenda driving NAFTA. Spanish-speaking women, who once migrated to El Paso in search of a better life, have had a rude awaking after dedicating their entire lives to building the city's economy. After having been subject to exploitation and limited employment in El Paso, these women are now being marginalized and slowly eliminated from an increasingly internationalized economy.

To mobilize displaced women workers, LMO has focused on Mexican culture as a vehicle that links individual members to the organization. The usage of Mexican religious symbols, traditions, histories, and customs has served to reinforce the bonds between the organization and its members. At the same time, its creation of a transnationally shared space of Mexican culture has enabled LMO to become a legitimate economic actor in the city's economic restructuring and to resist a transnational agreement that has negatively affected their lives. The success of this strategy has manifested itself in the fact that no other organization in the U.S.-Mexico border region has achieved as much as LMO. The LMO has helped obtain El Paso's designation as an Empowerment Zone, won a $45 million grant for displaced workers, created two umbrella organizations—the *Asociacion de Trabajadores* and *El Puente*—opened its own school for displaced workers, established a childcare facility, and is currently developing the first bilingual adult curriculum for displaced workers. All of LMO's achievements serve as a testament to the power of a transnationally based culture in women's political activism.

What kinds of lessons can be learned from the success of this particular movement at the U.S.-Mexico border for ongoing international efforts at grassroots globalization? A first lesson points to the fact that organizing around any one specific identity—in this case the Mexican national/ethnic identity—may not necessarily suffice to forge politically viable cross-border alliances. Other principal factors may come into play as well, such as gender, experiences, values, beliefs, ideologies, and so on. In addition, mobilization depends upon the way in which an organization chooses to construct or define its identity, given a certain set of circumstances. Whether or not that identity resonates with its members has a bearing on the perception that external agencies, individuals, and institutions have of that organization, which may have significant political repercussions, for example, on access to resources of governmental agencies.

Perhaps more importantly, this study reveals the emergence of themes that workers on both sides of the border could mobilize around, such as the pursuit for social justice among all workers and the push for "living wages." By participating in broad movements for social justice, the LMO would broaden its support base and become more inclusive of other grassroots organizations to catapult them into the international area where their voices would be more difficult to stifle. A successful example of such an international forum was the recent gathering of numerous grassroots movements at the Summit of the Americas meeting on April 20–22, 2001 in Quebec City to protest the expansion of economic integration (free trade) to include the whole Western hemisphere at the expense of social justice.

And finally, this essay also points to the growing participation of women—and women of color—as leaders, not just participants, in grassroots mobilization against NAFTA, the Free Trade for the Americas Initiative (FTAA) which is scheduled for completion in 2005, the Multilateral Agreement on Investment (MAI), and other similar agreements. In a recent article, Lisa Montoya and her colleagues point out that political scientists have tended to neglect or discount Latina leadership and participation in electoral and community politics. A key issue within the debate about gender differences is whether there is an essential divide between the public and private dimensions of politics. For Latina women, more so than for men, the boundary between these supposedly distinct spheres of life is often blurred. With their emphasis on grassroots politics, survival politics, and the politics of everyday life as well as with their emphasis on the development of political consciousness, Latina women see connections between the problems they face personally and the community issues that stem from government policies.

Notes

1. I would like to thank my parents for their unconditional support. This study would not have been possible without them.
2. This case study is based on fifty-six face-to-face interviews I conducted from August 1998 through August 1999 in El Paso, Texas with women who are active in LMO and have lost their jobs because of NAFTA, that is, businesses moving from El Paso to Mexico. I have defined as "active" those women who attend weekly meetings, belong to committees within the organization, volunteer their time, attend and help in organizing *fiestas* (parties), rallies, and protests. The interviews were conducted in Spanish (and on occasion in English) at LMO, protest and rally sites, and, when it was convenient for the interviewee, over the phone. I have also included interviews with the lead organizers of LMO, a former bilingual teacher, who was employed to teach these women English at one of the many schools that offer English classes, a Public Relations spokesperson from one of the last mammoth garment industries (Levi Strauss) in El Paso, politicians and various governmental/political officials. Depending on the individual, these interviews lasted between fifteen minutes to one hour and fifteen minutes. In LMO, the women speak only Spanish, range in age from 35 to 72, and more than half have no more than a fourth-grade education in Mexico. At the time of my interviews with the women of LMO, almost all of them were either in school learning English, studying for their Graduate Education Diploma, or in some retraining program.
3. For a more complete discussion, see Ronald Takaki.
4. According to Brecher and Costello, the term "race to the bottom" denotes the reduction in labor, social, and environmental conditions that result from global competition for jobs and investment. For a complete discussion on the "race to the bottom," see Jeremy Brecher and Tim Costello's *Global Village or Global Pillage.*
5. In my conversation with John Ownby of the Dislocated Worker Service Unit at the Texas Workforce Commission on September 1, 1999, I was told that being classified as a "certified NAFTA displaced worker" might not reflect the true number of displaced workers. Businesses that close up shop due to NAFTA are supposed to certify their workers with the Texas Workforce Commission. Certification by the Texas Workforce Commission ensures that the displaced worker will be eligible for retraining programs. However, the Texas Workforce Commission does not have the manpower to see that

businesses do, in fact, register their workers as certified NAFTA displaced workers. Thus, the official number of certified displaced workers may not be a true account of displaced workers. Ownby believes the number to be much higher.

6. This information is based on face-to-face interviews and informal discussions with women.
7. Guillermo Glenn, personal interview, January 27, 1999.
8. See Coyle et al. 117–143.
9. Maria Antonia Flores, personal interview, October 13, 1998. Author's translation of the original Spanish.
10. Glenn, personal interview.
11. Maria Antonia Flores, personal interview. Author's translation of the original Spanish.
12. The power of culture as a mobilizing force is not a new phenomenon. It has been instrumental in the Civil Rights movement, the Chicano movement, and the farm workers movement as well as many others.
13. Keith Klapmeyer, personal interview, November 19, 1998.
14. Ibid.
15. For more information, see Herrera-Sobek; Rodriquez.
16. Paz Orquiz, personal interview, January 20, 1999. In both my formal and informal conversations with LMO and it members, this was a response given by all.
17. Refugio Arrieta, personal interview, November 11, 1998.
18. Anonymous, personal interview, November 4, 1998.
19. A *corrido* is defined as a simple narrative ballad, which relates an event of interest only to a small region; it may be a love song or a comment on the political situation. For more information on *corridos,* see Dickey.
20. Cindy Arnold, personal interview, June 21, 1999. As a political actor, LMO emphasizes that it neither affiliates with one specific group nor endorses any particular political party or politician. Its legal charter as a community-based organization precludes involvement in electoral politics. Now that half of their members are U.S. citizens and thus eligible to vote, however, the organization will also encourage its members to vote.
21. Anonymous, personal interview, March 17, 1999. Author's translation of the original Spanish.
22. Irma Montoya, personal interview, February 24, 1999.
23. Lorenza Reyes, personal interview, January 13, 1999; for more information, see also Fernández-Kelly; Garcia.
24. This name has been altered to protect the identity of the woman who revealed this information in a private, informal conversation.
25. Arnold, personal interview.
26. Flores, personal interview.
27. Martha Ojeda, personal interview, January 21, 2000.
28. Ibid.; see also Alexander and Gilmore 44–49.
29. Ojeda, personal interview.

Works Cited

Alexander, Robin and Peter Gilmore. "Official and Independent Unions Angle for Power in Mexico," *NACLA Report on the Americas* 28.1 (July/August 1994): 44–49.

Ayres, Jeffrey M. *Defying Conventional Wisdom.* Toronto: University of Toronto Press 1998.

Barnet, Richard J. and John Cavanagh. *Global Dreams.* New York: Simon and Schuster, 1994.

Benedict, Anderson. *Imagined Communities.* London: Verso, 1983.

Brecher, Jeremy, and John Browns Childs, and Jill Cutler. *Global Visions: Beyond the New World Order.* Boston, MA: South End Press, 1993.

Brecher, Jeremy and Tim Costello. *Global Village or Global Pillage: Economic Reconstruction From the Bottom Up.* Boston, MA: South End Press, 1994.

Coalition for Justice in the Maquiladoras. "Duro Workers Intercept Fox's Tour in Reynosa," *Press Release* (January 12, 2001): 1 and 2.

Coyle, Laurie, Gail Hershatter, and Emily Honig. "Women at Farah: An Unfinished Story," *Mexican Women in the United States.* Eds. Magdalena Mora and Adelaida Del Castillo. Los Angeles, CA: Chicano Studies Research Center Publications, 1980.

Dickey, Dan William. *The Kennedy Corridos: A Study of the Ballads of a Mexican American Hero.* Austin, TX: The Center for Mexican American Studies and the University of Texas at Austin, 1978.

El Paso Greater Chamber of Commerce. *The El Paso Labor Market: A Training Gap Analysis, Final Report.* El Paso, TX: December 1997.

El Puente CDC/La Mujer Obrera. *Building Employment and Economic Development Bridges in South Central El Paso for Displaced Workers Strategic Plan.* El Paso, TX: La Mujer Obrera, 1999.

Fernández, Kelly, Maria Patricia, and Anna M. Garcia. "Power Surrendered, Power Restored: The Politics of Work and Family among Hispanic Garment Workers in California and Florida," *Women, Politics, and Change.* Eds. Louise Tilly and Patricia Gurin. New York: Russell Sage Foundation, 1990: 130–149.

Friedman, Susan Standford. *Mappings.* Princeton, NJ: Princeton University Press, 1998.

Gabriel, Christina and Laura MacDonald. "NAFTA, Women and Organizing in Canada and Mexico," *Millennium: Journal of International Studies* 23.3 (1994): 535–62.

Gilot, Louie. "Displaced Workers Want City to Help Find Jobs Quickly," *El Paso Times* (August 31, 1999): 10B.

Harison, Bennett. *Lean and Mean: the Changing Landscape of Corporate Power in the Age of Flexibility.* New York: Basic Books, 1994.

Herrera-Sobek, Maria. *The Mexican Corrido.* Bloomington, IN: Indiana University Press, 1990.

Hurrell, Andrew and Njaire Woods. "Globalization and Inequality," *Millennium: Journal of International Studies* 24.3 (1994): 447–470.

Kolence, Vic. "Apparel Industry was Precursor to Maquiladoras," *El Paso Times* (February 21, 1999): 4.

La Botz, Dan. "Reform, Resistance, and Rebellion Among Mexican Workers," *Borderlines* 6.7 (September 1998): 2–12.

La Mujer Obrera. *Stop NAFTA's Violence Against Women Workers.* El Paso, TX: La Mujer Obrera, 1999.

———. *Women Working Together: Goals and Activities.* El Paso, TX: La Mujer Obrera, 1999.

———. *Progress Report.* El Paso, TX: La Mujer Obrera, 1998.

Larudee, Mehrene. "NAFTA's Impact on U.S. Labor Markets, 1994–1997," *Pulling Apart: The Deterioration of Employment and Income in North America Under Free Trade.* Eds. Bruce Campbell, Maria Teresa Gutierrez Haces, Andrew Jackson, and Mehrene Larudee. Ottawa: Canadian Centre for Policy Alternatives, 1999.

Miller, Mike. "Citizen Groups: Whom Do They Represent?" *Social Policy* 22 (Spring 1992): 54–56.

Montoya, Lisa, Carol Hardy-Fanta, and Sonia Garcia. "Latina Politics: Gender, Participation, and Leadership," *Political Science and Politics* 33.3 (September 2000): 555–561.

Moreno, Jenalia. "Border City Struggles to Mend Loss of Apparel Industry," *Houston Chronicle* (December 12, 1997): 1A.

Mrkvicka, Mike. "Pushing For a Better El Paso: 10 Business Leaders to Shape City in '99," *El Paso Times* (January 10, 1999): 1E.

Nagel, Joane. "Masculinity and Nationalism: Gender and Sexuality in the Making of Nations," *Ethnic and Racial Studies* 21.2 (March 1998): 242–269.

Navarro, Marysa. "The Personal is Political: Las Madres de Plaza de Mayo," *Power and Popular Protest.* Ed. Susan Eckstein. Los Angeles, CA: University of California Press, 1989: 241–258.

Nesiah, Vasuki. "Toward a Feminist Internationality: A Critique of U.S. Feminist Legal Scholarship," *Harvard Women's Law Review* 16 (1993): 190.

Olvera, Joe. "Hunger Strikers Face New Battle on Home Front," *El Paso Times* (September 2, 1990): A12.

Pardo, Mary S. *Mexican American Women Activists.* Philadelphia, PA: Temple University Press, 1998.

Peterson, V. Spike. "Transgressing Boundaries: Theories of Knowledge, Gender and International Relations," *Millennium: Journal of International Studies* 21.2 (1992): 183–206.

Pulido, Laura. *Environmentalism and Economic Justice.* Tucson, AZ: University of Arizona Press, 1998.

Rodriguez, Jeanette. *Our Lady of Guadalupe: Faith Among Mexican-American Women.* Austin, TX: University of Texas Press, 1994.

Rosenman, Mark. "Response: A Place for All Nonprofits," *Social Policy* 22 (Spring 1992): 57–59.

Spener, David and Kathleen Staudt. *The U.S.-Mexico Border.* Boulder, CO: Lynner Rienner Publisher, Inc., 1998.

Takaki, Ronald. *A Different Mirror: A History of Multicultural America.* Boston, MA: Little Brown, 1993.

Templin, Neal. "Anatomy of a Jobs Program that Went Awry," *Wall Street Journal* (February 11, 2000): B1 & B4.

Transborder Collaboration

The Dynamics of Grassroots Globalization[1]

Manuel Rafael Mancillas

This article documents the work of the Border Arts Workshop/Taller de Arte Fronterizo (BAW/TAF) in the Poblado Maclovio Rojas, a citizens' organizations of "land squatters" near Tijuana, Mexico. Throughout the study, I examine the forces that have affected the long-term collaboration between the San Diego-based grassroots cultural workers' collective and the Poblado. The Maclovio Rojas-BAW/TAF partnership has taken its current shape within the larger context of a globalized capitalist economy that appropriates territories and breaks down national borders. Our collaborative project is one of hundreds, perhaps thousands, of other, similar transborder (Tijuana/San Diego border) partnerships.[2]

As I became aware of the current fashionable extension of "border studies" to also include questions of gender, culture, politics, age, and sexual identities as well as of north-south, east-west, and urban-rural boundaries, I decided to expand the study beyond a mere chronicle of crossborder collaboration between the BAW/TAF and the Poblado to also focus on other types of border crossings. The cooperation between the BAW/TAF and the Maclovio Rojas community has exposed and transformed several traditional boundaries, such as those between the artist/intellectual and the *maquiladora* worker, between the "first world" petit bourgeois and the "third world" proletariat, between art and politics, and between object and subject.

As a community of artists and cultural workers, the BAW/TAF has struggled above all with questions about the role of the intellectual within and outside of community based movements. Members of the BAW/TAF are engaged in the cultural production of what has often been dismissively labeled as political propaganda art. In our experience, however, the BAW/TAF has held its aesthetic quest to a higher level of criticism. BAW/TAF members are conscious that there is a difference between the use of politics as a prop and the use of artwork as a propaganda tool. In the latter case, art supports the struggle against power structures that oppress communities, artists, and intellectuals alike.

After his tour of the southeastern Mexican states, José Saramago, the Portuguese poet laureate, said that "the trip strengthened my belief that if it took humanity millenniums to build its conscience, [so] to not use it and go where there is suffering, where there is death, where there is humiliation, then, what is the use of conscience?" When the Swedish Academy honored Saramago with the 1998 Nobel Prize in Literature, it recognized the poet's commitment to the only possible history—the underground history of solidarity in the face of massacres, wars of extermination, and other tragedies. In the preamble to his acceptance speech, Saramago told reporters that "The world could be named Chiapas and everything would be very clear: suffering, misery, hunger, injustice, everything is there. If one has a mouth and a thought and a capacity to express it and one does not speak of this, then I believe that it would be more or less dead."

Frameworks

The Border Arts Workshop/Taller de Arte Fronterizo BAW/TAF was originally formed in 1985 for a group installation/performance project at La Galeria de la Raza in San Francisco in partnership with the Centro Cultural de la Raza in San Diego. The Workshop has traditionally functioned as a collective of artists and cultural workers, but it has also attracted nonartist collaborators throughout the years. BAW/TAF has no formal organizational structure and relies on outside support for the acquisitions of funds, primarily from the Centro Cultural de la Raza. In addition, the Workshop has received several grants from the National Endowment for the Arts, the California Art's Council, and the City of San Diego Commission for Arts and Culture.

The BAW/TAF has been invited to participate in the Venice Biennale, the South Africa Johannesburg Biennale, and the Habana Biennale. We took part in a six-month residency program at the Sidney Biennale, where we worked with Aborigines and Indochinese refugees. Since 1991, the Workshop has participated in several short-term community engagement projects—with Hallwalls in Buffalo, New York; with the Vista Elementary School District for the LACE "Destination LA"; with Palomar Community College in San Diego's North County for inSite '94; and with the Children's Museum/Museo de los Niños in San Diego. The latter work was awarded the first prize in the children-under-19 category for the 1996–1997 World

Ideas for Peace by the Habitat II, which was sponsored by the United Nations for the fiftieth anniversary of UNICEF.

The membership of the Workshop has changed throughout the years, and only Michael Schnorr, a professor of Art in the Southwestern College, Chula Vista, has remained of the original members. The BAW/TAF's most recent collaborative project with the Maclovio Rojas community constitutes an enormous shift in the type of projects the Workshop has been engaged in. My friend Arthur Himmelman, a national consultant on questions of collaboration,[3] has defined "collaboration" as

> the most complex process along a developmental continuum, that includes networking, coordination, cooperation, and collaboration. Specifically, and for the purposes of this paper, organizational collaboration is defined as the process in which organizations exchange information, alter activities, share resources, and enhance each other's capacity for mutual benefit and a common purpose by sharing risks, responsibilities and rewards. (Himmelman 1994)

The collaborative relationship between the BAW/TAF and La Unión de Posesionarios del Poblado Maclovio Rojas Márquez, A.C. has grown along the lines of the developmental continuum Himmelman describes. According to Himmelman, the New Global Order requires that communities and workers "must learn to do more with less" in the aftermath of neoliberal decision-making that "is specifically designed to reduce resources for human development and community revitalization not considered essential to global investors" (Himmelman 1996). Himmelman's work provides a map that has helped us to constantly evaluate and rethink our relationship with the Poblado. He writes that it is necessary to "extend our expectations [of doing more with less] to question issues of power, and the transformation of power relations through the collaborative engagement" (Himmelman 1996).

For the partnership between the Maclovio Rojas residents and BAW/TAF to develop it was very important to accurately define the nature of our collaboration every step of the way as it has moved beyond a site-specific, time-specific, and grant-driven partnership. The dynamics generated by the challenge of our project with Maclovio Rojas have also significantly transformed the make-up of the Workshop. Besides receiving support from many other artists and volunteers, the BAW/TAF members who participated in the collaborative project with Maclovio Rojas were Lorenza Rivero, Berenice Badillo, and myself. As a native of the Tijuana/San Isidro border region, I have become a master of transborder crossings, a *zopilote* who has wagged the tales of *la linea* and lived to tell you about it, perched unvacillating high on the fence.

Photo 9.1. Painting of Maclovio Rojas Márquez on the Stage of the Cultural Center in Maclovio Rojas. Painting by Yolanda Romero. Photograph by Michael Schnorr. Photograph courtesy of BAW/TAF.

Mistaken Identities[4]

I first heard about the struggle of the Poblado Maclovio Rojas residents during my daily routine of listening to the radio talk shows from Tijuana. One of the radio *tribunas* reported that a group of squatters were demanding the recognition of their rights to land that they been occupying for eight years. In response, the Urban Development officials of the State of Baja California tried to repress their petition with police force and threats of jail.

The Poblado Maclovias Rojas is a self-governed community of land squatters that is based on consensus and communitarian democracy. The community has its roots in the 1987 *posesionario* movement that was organized by the CIOAC (Central Independiente Campesina de Obreros Agrícolas y Campesinos), a national, independent, and democratic farmworkers and peasant organization. Formed in the 1940s by Ramón Danzós Palomino, a member of the old Mexican Communist Party, the CIOAC was the first rural-based organization to challenge the official *campesino* organizations, which were then largely controlled by the ruling party (the PRI) and the government. The CIOAC had a history of organizing migrant Mixteco farmworkers in the tomato fields in the state of Sinaloa, and in San Quintín, Baja California.

At the time of his death, Maclovio Rojas was the secretary general of the CIOAC in San Quintín. When the first 25 women occupied the land of the Poblado in 1988, laying out makeshift houses built out of wooden pallets and blue tarp roofs, none of them had ever personally met Maclovio Rojas. To remember Maclovio's death, the CIOAC named the land movement and two others in Baja California in his honor.

The Maclovio Rojas land takeover was not news to Tijuana, to the urban centers in Mexico or to the rest of Latin America. Accelerated migration from rural to urban areas as a result of rapid economic growth within a globalized economy has created thousands of other "Maclovios" throughout the world. Since 1929, when the first *posesionario* movement took over *La Colonia Libertad,* these popular movements have significantly shaped Tijuana's urban development. Millions have left what Northerners call the interior of Mexico to migrate to the United States, but many of those have remained in the border area.

With the rapid growth in the *maquiladoras* and in the service industries described in Ursula Biemann's contribution to this volume, many migrants become immediately absorbed into the border labor pool, thus creating a housing shortage and conditions for land squatting. Tijuana's growth rate during the 1970s reached an outstanding 14 percent annually. Thus, unlike the controlled urbanization patterns patented by the long-term planners of Los Angeles City and County, for example, Tijuana's urbanization has been marked by a series of land squats and government repression against the *invasores.* This chaotic pattern of growth in a geographically semi-arid region of steep ravines and canyons with limited water resources, with the geopolitical boundary to the north and the Pacific Ocean to the west, has led to the concentration of almost 90 percent of the population in the urban core.

The current struggle of the *pobladores* of Maclovio Rojas for their 197 hectares (600 acres) of land is rooted in a long and very complex history of agrarian reform and land tenure issues. Maclovio Rojas sits on the southeastern area of what once was *El Rancho El Florido.* Two sisters, descendants of a Greek immigrant named Yorba, claim to have the original title of the land grant, which was awarded to them between 1834 and 1842. Yorba raised cattle for export to California and planted thousands of acres for an olive grove. This *latifundio,* or large land grant, was part of the land that the Mexican Republican independent government, after its independence from Spain, awarded to high ranking military officers of the insurgent army and to European immigrants. The government sub-divided former mission lands into *latifundios* that primarily served as *rancherias* for cattle raising. After the U.S.-Mexico War and the signing of the Treaty of Guadalupe Hidalgo in 1848, most of the Mexicanos and Californios had to leave their *rancherias* in Alta California and settled south of the Tijuana River.

The *pobladores* do not see themselves as *paracaídistas* (literally parachuters or land squatters); instead, they argue that they are simply observing a mandate of the Mexican Constitution that allows the occupation of idle lands. The Agrarian Reform law of 1917 (passed after the Mexican Revolution) allows the granting of *eijidos*—parcels of land that cannot be sold, leased or transferred, but that are assigned in perpetuity to the *eijidatarios.* In 1991, the federal Agrarian Reform Department granted the *pobladores*' request for communally shared tenureship of the land they inhabit. An officially stamped invoice from the federal Agrarian Reform Department is pasted to the walls of the Poblado's assembly hall as proof that the Unión de Posesionarios del Poblado Maclovio Rojas Márquez, A.C., has paid the government 5,678.34 pesos (approximately $1,892.78) for the land. Recently, state officials have appraised the value of the real estate at $10.00 a square meter, making the disputed 197 hectares worth $197 million.

As a result of the neo-liberal market expansion under NAFTA, which began during the Carlos Salinas de Gortari regime, a new Agrarian Reform law was introduced. The law "liberates" the *eijidatario* and their families from government regulations and allows them to sell or lease their land at market value. Heralded as a victory for the poor *campesinos* to be finally freed from years of government paternalism and bureaucratic nightmares, the new legislation, however, merely gives legal shape to what had already been happening illegally for decades, and it provides the agro-industrial corporate growers with the power to accumulate more land. Salinas's reforms to Article 27 of the Constitution, in fact, provided a blueprint for the expansion of the *neo-latifundistas*—the new land holders—who are tied directly to the export sector of agriculture from what is often called America's winter gardens.

In 1988, Carlos Salinas lost in the Baja California polls to Cuahtémoc Cárdenas of the PAN. Although the Baja Californian electorate had voted for the leftist coalition led by Cárdenas, a year later Ernesto Ruffo, a neoconservative from the Partido Acción Nacional (PAN) became governor. Ruffo's first edict was to endorse the *no invasiones* campaign by warning people that he would no longer tolerate land occupations. Under his regime, the state and not the federal government would regulate land tenureship. The Ruffo government began dismantling the old network that had led many *posesionario* movements.

For the *posesionarios* of Maclovio Rojas the timing of their movement thus turned out to be unfavorable. Baja California officials would not recognize the *eijido* entitlements awarded to them by the federal government, and because of the changes in the Agrarian Reform law and in the mechanisms for the management of land tenure cases, the Maclovianos never received their entitlements. Instead, they were served with several criminal warrants and threatened with state civil procedures for squatting and illegal homesteading. The original lien holders, the Yorba sisters, as well as other people who claimed legal land ownership initiated these civil suits. By January 2000, the issue of the ownership of Maclovio Rojas had not been legally resolved: Both federal and state governments have refused to determine the tenureship of the 197 hectars.

Since I had personally met Maclovio Rojas in San Quintín several years before his death, I became intrigued by the situation of this community that was named after him. Maclovio Rojas was a very special man of strong convictions and faith who died as he walked on the corridor of power. I had a photograph that I thought was him, and I decided to "rescue his image" by painting a picture of him somewhere in the community. I had never heard of an Indian from Oaxaca other than two of our presidents, Juárez and Díaz, who had had a community named after him/her.

We approached Jaime Cota, a labor right's organizer in Tijuana and member of the Frente Zapatista de Liberación Nacional, FZLN, the civilian support group of the EZLN (the Zapatista National Liberation Army), about painting a mural of Macolovio in the Poblado's meeting hall. Members and sympathizers of the Frente had built this small structure and makeshift stage, called Aguascalientes, out of garage doors and recycled wood.[5] Cota took Michael Schnorr and me to the Poblado and introduced us to Hortensia Hernández and Artemio Osuna.

Our cooperation began as Hortensia Hernández, leader and president of the base committee, was looking at the tiny 35 mm slide I showed her. She liked the image and agreed with me that it looked like Maclovio Rojas. At any rate, she was sure that the person standing next to him was his brother. We had a long conversation about Maclovio Rojas and about the legal problems facing the community. The Poblado was not only being threatened by the change in federal law I mentioned above, but also by its proximity to the production facilities of Hyundai Precision Co. Hortensia explained that in 1993 the Korean *maquiladora* simply appropriated one hundred hectares of the Poblado for storage and the parking of their cargo containers, and was now threatening to take the rest of their land. Three years earlier, Hyundai had relocated one of their manufacturing plants to Tijuana's El Florida Industrial Park as part of a Korea-Mexico negotiation agreement that Salinas de Górtari had signed to attract Asian investment to the border areas. Consequently, the real estate surrounding the industrial park, which is next to the Poblado Maclovio Rojas, became highly coveted by land speculators. Development adjacent to Hyundai storage facilities now includes two PEMEX gas stations, a mini-market with truck stop, and a new furniture assembly *maquiladora* plant.

The image on the slide I showed Hortensia had a dark shadow around the eyes caused by the rim of the hat. Unable to clear it electronically, we needed another photograph to get better definition. By what turned out to be a series of lucky strokes, we met Maclovio's brother, Lucio Rojas, who has been one of the main catalysts for the organization of Mixteco-Zapoteca migrant farmworkers in San Diego's North County. When we approached him, he told us that his family was still living in San Quintín, and that he was going to travel there during the following Fourth of July holiday. We decided to visit and interview the Rojas family members to secure a better photograph.

Traveling on the transpeninsular highway the 200 miles from the border south to San Quintín valley along the scenic Baja California coast is at once beautiful and treacherous. This corridor of power is congested with state-of-the-art refrigerated cargo containers that are transporting fresh tomatoes and other fruits and vegetables to places in the United States, such as New Jersey, Michigan, and Ohio. The narrow two-lane highway has barely enough room for containers that rush past along in what seemed like two-minute intervals. The valley began developing high-yielding agro-industrial farming for export in 1980. As the Southern California suburban land rush was displacing farmland to the south, the fertile San Quintín valley became the yearlong supplier of vegetables to the north.

Agricultural expansion also required cheap farm labor. Expelled from their homelands by poverty, Mixteca Indians from the state of Oaxaca quickly moved to meet this demand. At the height of the harvest in 1985, almost 80,000 farmworkers were laboring in these tomato *maquiladoras,* while living in labor camps inside the grower's property. Maclovio's family had immigrated here in 1980; he joined them in 1984. By 1986, he had become a leader and president of the Central Independiente de Obreros Agrícolas y Campesinos (CIOAC), a national organization that was unionizing farmworkers. As many leaders before him, he was faced with an enormous task since there has never been an independent union of farmworkers in Mexico. In the end, Maclovio gave his life for this cause. On July 4, 1987, he was run over by a truck as he crossed a highway. We now know that a grower ordered his murder and that a rival Mixteca leader carried out the killing.

As I was presenting the photograph to Maclovio's older brother José, he paused for a long time. Trying to find words, he politely thanked me for my good intentions and said that, unfortunately, it was the picture of his uncle Fausto. However, the other person in the photograph was his brother Lucio. The embarrassment was eased when the family kindly provided us with the only photograph they had of Maclovio, a photo taken on the day of his marriage.

Art in the Context of Collaborative Empowerment

After our return from San Quintín, a series of events expanded our project beyond the original, largely aesthetic idea of painting a mural of Maclovio's image on the stage of the Aguascalientes into a long-term collaboration with the community in Tijuana. First, Baja California State Judicial police arrested Hortensia, Artemio, and Juan Regalado. Hortensia was apprehended on trumped up charges of illegal possession of property and damages to private property. We immediately traveled to Tijuana's state government offices to document the protests by the residents of Maclovio, supporters from other communities, and several of Tijuana's labor and human right's activists. The protesters were hoping that the issue of the arrests would be resolved in Tijuana and that they would be able to avoid travelling to the State's capital in Mexicali 120 miles away. As the local representatives of the governor of Baja California failed to address the issue, the leadership resolved to march to Mexicali on September 4, 1996 to demand the freedom of the three compañeros.

On Wednesday morning, September 4, 1996, the Aguascalientes in the main plaza of the Poblado Maclovio Rojas was full of people. Women and children were milling around, painting banners and signs, preparing their bodies and souls for the road ahead, packing food, water, and hydrolyzed serum (used to fight off dehydration), which was donated by supporters. They wanted to march to meet with the governor of Baja California. Highway 2 would take the marchers through the 5,500 feet Sierra Juárez pass, down the Rumorosa grade to the Laguna Salada 110 feet below sea level, where temperatures can climb to 115 degrees at midday. Over 300 people began the march. Soon, impatient horns blasted through the morning sun, a massive traffic jam backed up for miles, and dirt and smoke filtered the colors flying in the sky. One marcher, Rubén Hernández, died while crossing the desert. The Maclovianos pledged to return a year later to the place where he died and to erect a monument in his honor.

Both of these events—the arrests of the Poblado Maclovio Rojas leadership and the protest march—considerably altered the shape of our project. The BAW/TAF not only video-documented the community's struggle in order to create a public record, but also provided direct support to the marchers. Through the BAW/TAF's network in Southern California, we are also able to mobilize broader support. We contacted support groups and several NGOs in San Diego, primarily the American Friends Service Committee (AFSC) and their local U.S.-Mexico Border Program, who then contacted others who struggle for transborder social justice and solidarity.

The Support Committee of Maquiladora Workers (SCMW), a non-profit NGO based in San Diego with a long history of supporting *maquiladora*

Photo 9.2. Mural Wall of the Cultural Center in Maclovio Rojas, depicting people in the March for Liberty, Summer 1997. Painted by Berinice Badillo, Michael Schnorr, Susan Yamagata, Glen Knowles, and Lewis Perez. Photograph by Michael Schnorr. Photograph courtesy of BAW/TAF.

workers in Tijuana,[6] developed a letter writing campaign to demand the freedom of the compañeros.[7] In addition, an article by Julio Laboy in the California section of the *Wall Street Journal* from February 2, 1997, demonstrated to corporate investors the strong U.S. support for the struggle in Maclovio Rojas by listing the Poblado's many "friends within the belly of the monster." After the article's publication, Hyundai removed the cargo containers that were stacked next to the Poblado, and today only a few hundred of them remain in the lot.[8]

The article, however, also misled readers by linking Hortensia Hernández to the political arm of the EZLN as a caption below her image referred to her as "Sub-Comandante Hortensia." In a later press conference conducted in the Poblado, Hortensia Hernández made it clear that while the residents of Maclovio Rojas support and identify with the EZLN struggle and that of the indigenous communities in Chiapas, the Poblado's organization does not represent a political arm of the Zapatistas. In fact, there is no such official "political arm" of the EZLN.

The Artist as a Vehicle for Community Action

Since these events, the role of the BAW/TAF in the Maclovio Rojas community has been constantly redefined. During its annual "Border Realities XI" installation and a performance at the Centro Cultural de la Raza in San Diego, the workshop constructed an installation room that depicted the struggles of Maclovio Rojas. Members of the Poblado's base committee spoke at the event, and, in the following week, we were invited to begin discussions about a long-term community project in Maclovio.

Because of its participation in the inSite '97, a triennial transborder public art festival, BAW/TAF received funding for a community engagement project.[9] By securing additional funds from the U.S.-Mexico Fund for Culture, we were able to commit to a long-term project with Maclovio Rojas. BAW/TAF began negotiations with the community's leadership about the context of our participation in the community. We first initiated painting workshops for children, which took place every other week in the parking lot of the Poblado's assembly hall. After the community asked for more frequent visits, we requested storage space for our materials and equipment.

Instead, the base committee decided to provide us with the space for a new building, which would house the outreach offices of several regional organizations. The BAW/TAF presented its design of a two-story structure to be made from discarded wooden garage doors. In the process of locating these doors, we discovered another apt manifestation of grassroots mobilization. In the *Pennysavers* we noticed ads promising the installation of new aluminum fold-roll up garage doors and the removal of the old wooden

Photo 9.3. U.S. Garage Doors being moved to Maclovio Rojas. Photograph by Michael Schnorr. Photograph courtesy of BAW/TAF.

garage doors. When we contacted these garage door companies, they told us about crews of pirates from Tijuana that would steal the doors from their stores and take them across the border, thus saving the company the trouble and money of discarding the doors through the local refuse system. These *piratas,* would travel throughout San Diego County, scavenging for discarded garage doors through the back allies of suburbia. By crossing the border, American garage doors were transformed into walls for the houses of squatters. As these doors became incorporated into homesteads in Tijuana, they enabled the *posesionarios* to avoid dislocation. Many of the houses in the Poblado Maclovio Rojas were made of similarly discarded garage doors and wood pallets; however, newer houses are now being erected out of cinder block and mortar.

The dynamics of grassroots globalization works out that way, the *rasquachi* way, as rag-tag bands of border pirates transform local concerns into global affairs.[10] A globalized grassroots consciousness is similarly manifested in transborder alliances, partnerships, and collaborations, which transform *rasquachi* bands of disenfranchised community artists and political activists from the lunatic fringe into major players in the struggle for survival and the production of power.

Long Term Commitments: Breaking the Rules

The base committee decided to build the *foro cultural* out of cinder blocks to ensure more longevity and security for the building. Therefore, the wooden garage doors, which we had originally acquired to build the center, were instead used to line the perimeter of the area. We painted them with murals depicting the community's struggle and history and, in the process, also accomplished our original intent of painting the image of Maclovio Rojas at the top of the Aguascalientes stage area.

It took the Workshop exactly one year from the time of its initial negotiations with the base committee to finish the construction of the center. The building, called Aguascalientes, was inaugurated on July 4, 1998 to commemorate the eleventh anniversary of Maclovio Rojas's death. Soon thereafter, BAW/TAF presented the leadership of the Asociación of the Poblado with a preemptive document, responding to a request from the community for a "cultural plan." The document intended to formalize the collaborative relationship between the two partners and to create a contractual agreement that would serve as the foundation for long-term commitments.

The document called for the creation of a *Comisión de Cultura,* a board of directors of the Aguascalientes to be made up by community members. It also stated that the BAW/TAF would pay the salary of the staff to coordinate and promote the activities of the Aguascalientes. The BAW/TAF suggested

Photo 9.4. Girls in Front of House Made of Garage Doors. Photograph by Michael Schnorr. Photograph courtesy of BAW/TAF.

Photo 9.5. The Cultural Center in Maclovio Rojas. Photograph by Michael Schnorr. Photograph courtesy of BAW/TAF.

that the *Asociación* select the staff from the community and appoint volunteers to serve in the *Comisión de Cultura.* BAW/TAF requested that the second-story space of the Aguascalientes would be designated for the use of the Workshop (for storage and residency) and that the bottom floor would be set aside for art workshops and classes. Despite several attempts to develop such a cultural plan, no formal agreement between the Poblado and the BAW/TAF has been reached; the *Comisión de Cultura* has remained an idea, and plans for the Aguascalientes continue to be made unilaterally by each of the partners.

In the meantime, in spite of jails, attacks, threats, and divisive actions undertaken by the government against the Poblado, the Maclovianos are developing an infrastructure for their community and do so independently of the government. While community-based planning is a phenomena that has become commonplace in Southern Californian cities over the last forty years, in Tijuana urban planning and regulated development has only been occurring in the last decade. After years of having no electricity, the Pobladores have created webs of lines that criss-cross the streets, streaming down from the power poles that the government has installed on the street but not connected to their homes. After years of being without water, they have tapped into an aqueduct that carries water through the middle of their neighborhood to the nearby Samsung megaplant and have built a web of garden hoses to take water to their homes.

With support from their national organization, CIOAC-Democrática partners, the community is currently developing an economic plan that calls for *granjas familiares* (family farms) and a regional farmer's market to house *tiangüis* or *mercado sobre ruedas* (swap meets). The present swap meet fills up five blocks during Wednesday and Saturday mornings. The prized roadside commercial lots are being developed with small family operated businesses. The community's La Casa de la Mujer will soon open a large extension to their day-care center, including a nascent arts and crafts cooperative that is organized by BAW/TAF collaborators.

To anchor its economic development, the leadership of Maclovio Rojas has created a community bank to provide small interest loans from the money generated by the association's membership fees and monthly dues. The leadership decided to place the bank in the first floor of the Aguascalientes community center, thus occupying a space originally designated for cultural workshops. Consequently, the Aguascalientes was split in half, one part housing the bank and the other continuing to provide space for meetings, workshops, and classes.

The community leadership's unilateral decision created a major roadblock in the collaboration between the BAW/TAF and the Poblado. Members of the Workshop felt that the trust, developed in three years of collaboration had been broken. The majority of the members of the BAW/TAF thought that it would be better to build another building for the bank adjacent to the Aguascalientes. The decision by the leadership was made because of the immediate availability of the Aguascalientes and for reasons of internal political security. Poblado leaders argued that the Aguascalientes was underused as a cultural space and that, overall, cultural activity was not developing as expected, even though, at the time, several workshops and English classes were under way. Weekly ceramic workshops were held by BAW/TAF collaborators in an outside temporary space equipped with wheels and two kilns. BAW/TAF engaged in other activities, including the never-ending detailing and upgrading of the building, of the stage, and of other areas of the Aguascalientes. This level of cultural activity was apparently not meeting the expectations of the community's leadership for BAW/TAF, but the expectations had not been made clear and lacked the foundation of a cultural plan. Whether or not a written contractual agreement about the collaborative partnership would have made a difference in the shaping of the relationship between the BAW/TAF and the *Asociación* remains unclear. A transformation of their relationship is more likely to happen in praxis, even though a basic agreement could have established a basis for better accountability.

Despite this problem, the BAW/TAF has made a long-term commitment to the partnership with the community and is now a member of the Poblado's general planning committee. The BAW/TAF realizes that it is the committee's prerogative to set priorities for their community and to control its resources. Members of the BAW/TAF now join the Poblado sector's delegates, block captains, and the lead individuals in the planning of each of their infrastructure projects. BAW/TAF is currently the only organization outside of the Poblado that participates in this committee.[11]

The partnership continues to advance today, mainly because both sides are satisfied with the current arrangement. The community receives programming by artists and cultural workers, and the BAW/TAF gets the benefit of a community willing to accept them and to provide a basis for future projects and funding. The BAW/TAF was recently awarded the second prize for an installation about the Maclovio project in an exhibit in Tokyo. Collaborators from Argentina, Australia, France, Sweden, Brazil, San Francisco, Los Angeles and other local artists have come or have stated their willingness to participate in a residency project in the community. The Orange County Friends of Maclovio Rojas, Global Exchange, graduate students from Claremont College, and the American Friends Service Committee (AFSC), among others, have sent working delegations and/or have provided in-kind donations. The BAW/TAF continues to seek additional funding for the expansion of the project.

Crossing Borders

Although the BAW/TAF has come a long way in advancing the collaborative partnership with the Pobladores of Maclovio Rojas, much work remains to be done toward attaining the goal of power sharing between each of the partners. During the project, we have crossed many borders and we have traversed the labyrinth of several corridors of power established by globalization. Every border crossing has a power tariff that must be dealt with, either by confrontation or collaboration. Arthur Himmelman reminds us, however, that "the hope for communities all over the world is to build a new democracy through these collaborative processes, to transform power relationships where power is shared equally to enable us to mature into the so-called global village." All the power to all the people!

Notes

1. This work was first presented at the American Studies Association/Canadian Association of American Studies Conference in Montreal; and, at the "Cultural Studies and Disciplinary Boundaries in Latin/o America," at the University of Illinois, Urbana-Champaign in 1999. My thanks to inSite '97 and to the Fideicomiso para la Cultura México-Estados Unidos (U.S.-Mexico Fund for Culture) for the initial funding of the collaborative project. Special thanks to my compañera Susana Martinez and mis hijas, Yax-Ha, Marla, and Anya, for all their support and patience. To Greg "Pájaro" Bird for all of his webmastering tutoring and guidance. To Claudia Sadowski-Smith for believing in this work. To BAW/TAF and all the collaborators and friends that have helped along the way. To name some of them: David Harding, Rose Costello, Jorge Hinojosa, Jaime Cota, Carmen Valadez, Tanya Aguiñiga, Sofía Sánchez, Mary Ann Murname, Monica Nador, Martín Weber, Allesandra Sanguinetti, Xavier De La Vega, Nadine Picard, Peggy Kala, Rebeca Rivero, Brenda Montero, Mayumi Shimizu, Kish Stjerne, Belinda Valdez, the American Friends Service Committee-Western Region Youth Programs, Maylin Martinez, Alberto Caro, Yolanda Romero, Olga Odgers, Glen Knowles, Félix Pérez, Susan Yamagata, Todd Beattie, Leticia Jimenez, Roberto Martinez, Nadia and Diana Hernández, Lupita Guerrero, Judi Nicolaides, Jeff and Carolina Pilch, Paul Vaughn, Orange County Friends of Maclovio Rojas, Elizabeth Cazessus, Jorge Peña, Elías Rámirez (Trago Amargo), Waldo López, Taige Diver, Jason Beedle, Juan Pazos, Christy Funk, El Campo RUSE, Gary Ghirardi, Ben Tú, Danza Mixcoatl, Supersamba, Wholesale Lumber Associates, Karl Hanson, Antonio del Rivero—CLON, Sophia McClennen, Ron Strickland, the Marxist Literature Group (MLG), the Global Exchange-Discovering California Youth Program, the John Wolman High School in San Francisco, Pitzer College, Kristen Aliotti, Denise Moreno Ducheny, Southwestern College. To José "Chino" Carrillo

and most of all, to the Maclovianos for their strength in struggle. Su lucha es nuestra también.

2. A more in-depth analysis of transborder relations will require an extended research project to accurately provide information about all the institutions and organizations involved.
3. It is important to note that the word "collaborator" took on a different meaning during the past wars, i.e., then it meant working with the enemy and collaborating with the state.
4. I have written two articles "Corridors of Power" (*San Diego Review,* October 1996) and "(with-in/out) a Corridor of Power" (*Mediations* 1998) that provide more detail about the Poblado Maclovio Rojas. The articles introduce Maclovio Rojas Márquez the organizer-martyr, the community of Maclovio Rojas, and its protracted struggles.
5. The Aguascalientes in the Poblado Maclovio Rojas was built in the spirit of the EZLN's Aguascalientes. Currently, there are five Aguascalientes in Chiapas. The insurgent army built the Aguascalientes with the mission to serve as links to civil society and places where a culture of resistance could develop. This *sociedad civil* links a local and national democratic movement to global struggles and appeals to indigenous peoples, students, workers, community associations, gays and lesbians, *barzonistas* (bankrupt native mid-range commerce and industrial entrepreneurs who suffered from the Peso financial crisis), old school leftists, new age rock stars, and many others. During their national campaign to connect to the Mexican civil society, an EZLN delegation also met with the Maclovio Rojas community in the Aguascalientes. A highly significant event and piece of border art took shape during this visit. As the delegation could not obtain visas or otherwise cross the border, a demonstration event was organized along the 12-foot fence, in an area near the crossing gate of transborder NAFTA trains. Scaffolding was set up to install a platform high enough for supporters on the other side of the fence could see and hear the EZLN speakers.
6. The support committee was founded in the early 1990s by activists from the Partido Revolucionario de los Trabajadores (PRT), Solidarity, and other individuals on both sides of the border. Originally known as the Committee to Support Maquiladora Workers, the organization publishes a transborder newsletter called *Fuera de Linea.* The newsletter, a coordinated effort including the University of California in San Diego student newspaper, *New Indicator,* aims "to inform and provide analysis to activists confronting the corporate economic integration on the U.S.-México border."
7. For many years, the SCMW has solicited resources to maintain full-time organizers in Tijuana. In addition, financial resources were distributed to the compañeros in Maclovio Rojas in support of their legal defense fund. The SCMW kept providing direct support by organizing fundraising NAFTA tours in the Poblado. Busloads of activists from the Southern California region visited Maclovio for fact-finding activities to discover the effects of NAFTA in the border region. Currently, the SCMW continues to have close

ties with the AFL-CIO and other U.S. labor organizations that have opposed NAFTA.

8. After this success, the SCMW turned its attention to supporting the efforts of the Han Young workers working to organize an independent union. Han Young is a subsidiary of Hyundai Corporation in Tijuana; it subcontracts from Hyundai to weld parts of the chassis of the cargo containers. Several of the original leaders of the Han Young workers, who began the organizing effort, were residents of the Poblado Maclovio Rojas. There is ample documentation of the Han Young worker's struggle, which is available through many publications, specifically *Z Magazine* and articles by freelance writer David Bacon.
9. The inSite triennial festivals are funded by several institutional governmental sources that are funneled through Installation Gallery, including FONCA (National Fund for Arts and Culture), CONACULTA (the national Council for the Arts and Culture), and INBA (the National Institute for the Fine Arts). The festivals are organized with the participation of many of the regional IGOs and NGOs as well as regional galleries and museums
10. For an excellent definition of *rasquachismo,* see Tomás Ybarra-Frausto.
11. Recently, the Poblado's Association has joined a statewide network called ENFOCA—Enlace Frente Obrero Campesino—a labor, farmworker and neighborhood associations network). This network is formed by the Sindicato 6 de Octubre (the labor union formed by the ex-Han Young workers), the Procuradoria de Derechos Indigenas (the Indigenous Rights Advocacy Committee), several farmworker organizing committees from San Quintín, and about ten statewide neighborhood associations across the state of Baja California. ENFOCA will provide legal consultation, advocacy, and organization support.

Works Cited

Himmelman, Arthur. "On the Theory and Practice of Transformational Collaboration," *From Social Science to Social Justice: Creating Collaborative Advantage.* Ed. Chris Huxham. London: Sage Publications, 1996: 19–43.

———. "Communities Working Collaboratively for Change," *Resolving Conflict: Strategies for Local Government.* Ed. Margaret S. Herrman. Washington, D.C., 1994: 27–47.

Ybarra, Frausto. "Rasquachismo: A Chicano Sensibility," *Chicano Art: Resistance and Affirmation, 1965–1985.* Los Angeles: University of California Press, 1988: 155–162.

Encounter with a Mexican Jaguar

Nature, NAFTA, Militarization, and Ranching in the U.S.-Mexico Borderlands

Joni Adamson

Any mention of ranching on the U.S.-Mexico border today is likely to touch off fierce debates over armed vigilante cowboys and their friends in the KKK rounding up undocumented Mexican immigrants who are reportedly delivered to the U.S. Border Patrol or, as some suspect, murdered. In several, widely reported press conferences that have drawn attention to the southern Arizona border region, rancher Roger Barnett and his brother Donald have claimed that they and a loose coalition of ranchers and white supremacists from all over the United States are responsible for having "caught" over two thousand immigrants. According to the Barnett brothers, these coalitions are "prepared to kill Mexicans" who are damaging his and other ranchers' property and causing harm to the environment.[1]

For the last four years, I have been collecting the folklore and oral histories of ranching families who live in the same region now made infamous by the Barnetts—the high-desert grasslands on both sides of the Peloncillo Mountains, which run along the border of Arizona and New Mexico and join the Sierra Madre range in Mexico. These ranchers unequivocally disassociate themselves from the outspoken brothers who, rather than support themselves primarily through ranching, engage in a lucrative motor vehicle towing business that is often in the news for its illegal practices.[2] The brothers' inflammatory rhetoric and their actions have, in fact, undermined much of the ranchers' hard work to restore ecologically degraded rangelands and to change negative attitudes about supposedly ecologically recalcitrant, predator-killing ranchers. Many of the ranchers fear that the actions of the Barnetts could give rise to even greater violence if immigrants, knowing that they are being pursued by the Border Patrol and hunted by U.S. ranchers, decide that they must take defensive actions when they encounter a man in a cowboy hat, perhaps on a horse, perhaps herding cattle.

In discussions of these tensions, ranchers often pause thoughtfully and add that the Barnetts are drawing national support from white supremacist groups and even some locals, in part, because of the widespread property damage and tons of debris left strewn through the arroyos and pastures of every ranch adjacent to the international border or situated along well-traveled migration routes into the United States. Undocumented Mexican, Latin, and South American immigrants are crossing the border in record numbers, and in their wake, they leave behind damaged fences, broken water tanks, scattered water jugs, empty food cans, plastic food wrappers, diapers, grocery bags, clothes, and blankets.[3] Ranchers report that when left scattered on grazing lands, this refuse can be ingested by livestock and cause their deaths. Since they are running small, family-owned farms, the continual repair of fences, the collection of trash, and the disposal of dead livestock strains ranchers' already thin financial resources and takes away from their time to perform other tasks essential to keeping their ranches operational.

Many of the ranchers see the property damage caused by undocumented immigrants as a threat more serious to the ranching industry than other issues that they have been facing for years. Ranching has been weakened by battles with large meat packing corporations, which keep stock prices low by importing foreign beef, and with mainstream environmentalists, who insist that ranchers be put completely out of business for the damage livestock has caused to public lands over the past several decades. The Barnett brothers claim to be concerned for the environment and characterize the clash between ranchers and undocumented immigrants as an us-against-them crisis. I will examine their claims near the end of this essay, but in the discussion that follows, I would like to put this conflict into a wider context by exploring the role that the North American Free Trade Agreement (NAFTA) and the U.S. Border Patrol have played in escalating the tensions between ranchers and immigrants on the U.S.-Mexico border.

I will focus my discussion by drawing some parallels between two groups that do not, at first glance, seem connected: The first are Mayan *eijido* farmers, many of whom joined the Zapatista National Liberation Army (EZLN). The EZLN burst out of the great Lancandon Rainforest in southern Mexico and took over three sizable towns in the state of Chiapas on January 1, 1994, the day that NAFTA went into effect. The second is the Malpai Borderlands Group, a small organization of southern Arizona/New Mexico ranchers—unconnected to the Barnett brothers—who have formed an alliance with the Nature Conservancy, the U.S. Forest Service, and university grasslands scientists in order to respond to the threats posed to the ranching industry by both economic forces and mainstream environmentalism. I will examine member Warner Glenn's account of an encounter with a rare Mexican jaguar and survey the ways in which the Malpai Group is working to improve the ecological health of their land by encouraging the growth of indigenous species and discouraging "alien" species. This discussion will show how both the ranchers and eco-political groups like the Zapatistas are confronting false dualisms that oversimplify the politics currently being played out over definitions of nature. I will argue that, put in this context, it becomes possible to see how NAFTA, which supposedly opens the borders of Canada, the United States, and Mexico to economic integration, is not necessarily contradictory to the current militarization of the border that is creating a "sacrifice zone" where ranchers and so-called illegal aliens clash, and where it is considered acceptable to discount the people and the environments they inhabit.

The Zapatista Rejection of NAFTA

Around the globe, states, corporations, and social movements have never been as responsive to environmental imperatives as they are now. Nature has taken a front and center position in transnational debates on land reform, free trade, and global scientific endeavors. Ecological politics is on the top of many people's agendas, thrust upon even the most recalcitrant national, cultural, and social groups because of the monumental problems that have arisen from its negligence. For example, in order to "fast-track" passage of NAFTA, the Bush Sr. administration reluctantly responded to mounting environmental concerns and agreed to carry on parallel negotiations concerning environmental issues alongside trade negotiations with Mexico and Canada (Goldman 293). Consequently, NAFTA is the first major trade agreement that somewhat includes the notion of international environmental obligations.

With the increasing pressure to view good ecological politics as key to global ecological health, it has become common to hear policy makers, corporation heads, scientists, environmentalists, and even corporate business leaders refer to such concepts as the "global ecological commons," the effects of ecological degradation on the "global citizen," and "global science." But those who talk about global ecological "commons areas" are not always the people indigenous to a given commons area, and their knowledges, ethnotechnologies, and material needs are often dismissed in favor of global ecological health. This situation creates disturbing political shortcuts that lead to troublesome social and ecological problems for both human and non-human species. For example, when elite Northern scientists refer to Central and South American rainforests as global commons areas—or the "lungs of the world"—and the World Bank fences off large areas of "pristine" rainforest as "biodiversity sites," they pit the rights of those who dwell in the rainforests, perhaps burning stands of trees to clear land for agriculture, against the rights of North American metropolitan populations who need the clean air produced in the rainforest.

NAFTA presents us with other examples of the immediate and often dire impact of global environmental imperatives on local people and places. Touted as the first environmental trade agreement, the implementation of NAFTA has, in the case of Mexico, led to widespread economic impoverishment of indigenous peoples and to the degradation of the environments they inhabit. In preparation for the passage of the global trade agreement, Mexico sought out foreign investment capital from global care-taking institutions such as the World Bank (with its professed environmental imperatives) in the 1980s. Mexico then implemented measures of economic restructuring that allowed, among others, the depletion of the Lancandon Rainforest in Chiapas, a global treasure that was too valuable to be left in the hands of its residents—mostly uneducated peasants. Since 1982, the rainforest has been cut down at the rate of 3.5 percent a year, falling to the international market for mahogany and making way for highways, resettlement, oil drilling operations, even airstrips for drug traffickers. Just 30 percent of the original 5,000 square miles of the rainforest remain.[4] With their exclusion not only from the remaining rainforest that once provided the basis of their subsistence economy but also from the wealth being extracted from their state without any benefit to them, residents of Chiapas have been falling further into poverty.

Then, in 1992, peasants also learned that they would no longer be guaranteed the few land gains they had made because a reform of Article 27 of the Constitution permitted the privatization of land held that had been held in non-alienable corporate status as communal land (*eijido*).[5] This reform meant that the government was free to expropriate Indian communal lands for the needs of national and multinational industries As Manuel Mancillas describes in more depth in his contribution to this volume, while the practice of *eijido* lands has roots in a time before the arrival of the Spanish in 1519, it was the Agrarian Reform Act of 1920 that formalized the formation of communal farms as collective entities with legal stature, specific territorial limits, and representative bodies of governance. These *eijidos* were created to satisfy the demands of peasants who had seen their communal village lands divided up by large agricultural estates and/or who served as laborers on these estates.[6] In Chiapas, however, entrenched resistance to land redistribution led to the formation of fewer *eijidos,* and the primary source of *eijido* land was opened through colonization of unused forested areas of the jungle.

The reform of Article 27 affected half the land in Mexico and, together with NAFTA's free trade policies, forced subsistence and small-scale farmers to compete with highly subsidized U.S. agricultural products in the Mexican markets (Stephen 80). In Chiapas, this meant that fewer indigenous peoples could support themselves by farming and that increasing numbers of people were forced either to seek employment in the large cities of Central and South America or to risk crossing into the United States in search of livable wages. But in 1994, Mayan *eijido* farmers dramatically rejected economic restructuring and NAFTA by staging a rebellion that had been in the planning stage for a number of years and that began on the day the trinational accords took effect.[7] Non-Mayan Subcommandante Marcos as one of their most visible spokespersons for efforts to seek international support issued a series of communiques, which linked changes in Mexican land reform to the North American Free Trade Agreement.[8]

The Zapatistas not only refused to legitimate the state's proposed agrarian reforms, they also supported the reclamation of communal lands, insisted that Article 27 be returned to its original intent, and, most importantly, called for inclusion of indigenous producers in all political and economic decision making (Stephen 90). They were not fighting to preserve the pristine rainforest for elite "global citizens" or longing for a mythical past where indigenous farmers engage in low-tech subsistence farming on communal lands. Rather, the Zapatistas demanded that the Mexican government provide them with the kinds of conditions that would allow them to become competitive in the global marketplace. Many credit the Zapatistas' demands for setting off a chain of events that would reveal the corruption in the Salinas government and force Mexico to loosen the purse strings for indigenous farmers. At initial peace talks, the government promised to build new houses for corn farmers in Chiapas and improve infrastructure in the poorest Indian communities (Simon A11).

Chiapas presents us with a clear case of the view modern states, multinational corporations, and global institutions have on "nature," which is often narrowly defined as "pristine rainforest or wilderness" regions teeming with biodiversity. While these institutions place "nature" at the center of transnational debates on land reform and free trade, they also look the other way as whole areas within these regions and the indigenous people who may inhabit them are sacrificed to economic restructuring and profit-making. Global, state, and corporate entities thus practice a contradictory politics that draws lines of protection around some areas while writing off others as sacrifice zones—defined as those places allowed to be logged, drilled, mined, or toxically contaminated for "the good of the national or global economy."[9]

In his study of the political resurgence of the notion of global commons areas, sociologist Michael Goldman points out that at the core of any ecological politics, be it radical or neoliberal, is the issue of proprietary rights to nature—who owns, manages, speaks for, and gets to make decisions about the future of "nature" or "wilderness." In this debate, false dualisms are imagined that muddle the politics being played out over environments. "North is pitted against South, global versus local, economic growth versus subsistence, abundance versus scarcity and rational science versus irrational [folk] beliefs [and practices]" (Goldman 13). These dualisms, then, become the apparati of institutions such as the World Bank, World Trade Organization, or NAFTA, which often erase the significance of local knowledges, specific places, and particular groups of people in favor of global economic objectives.

Groups like the Zapatistas enter the struggle over proprietary rights to their local environments by making it quite evident that ecological politics is not a singular, globalizing discourse. The way the World Bank might define "nature" or outline proper management of "global ecological treasures" may not be the way that the Mayan peoples who have inhabited the Lancandon Rainforest for millenia, deriving their livelihoods from it, may define it or outline it. The Zapatistas' struggle, then, is as much over what counts as "nature" and who gets to speak for the environments in which they live, as it is a struggle against racism, poverty, and authoritarian political systems. They insist that local communities of people have the right to be a part of any economic or environmental decision making that concerns them or the places they live and to also assert their right to derive their livelihoods through healthy and socially just forms of land use.

The Malpai Borderlands Group

While the effects of global economic objectives and the ecological imperatives inscribed into World Bank practices and NAFTA have already been explored with respect to groups like the Zapatistas (in part, because the group itself has articulated the problems so dramatically), other less well-recognized eco-political groups are also entering the struggle over global economics and proprietary rights to nature. The Malpai Group is another community that had to respond to NAFTA's economic challenges and to dominant definitions of "nature," "wilderness," and "proper" land management practices.

The Malpai Borderlands Group is a small alliance of some 35 ranchers who have come together to work for the preservation of two things that are widely construed as contradictory: a vital ecosystem and the preservation of the cattle ranching way of life. For many decades, members of the Malpai Group—and their fathers—have seen their livelihoods threatened by several forces. These threats include economic trends, federal managers who insist that ranchers and farmers do not have the scientific knowledge or technical know-how to make the best decisions about their own lands, and U.S. environmentalists who call for an end to ranching altogether because grazing threatens wild plants, animals, and ecosystems. More recently, these ranchers, living on both sides of the southern Arizona/New Mexico borderlands, have been faced with the dramatic increase in the numbers of undocumented immigrants crossing into the United States, in large part, because of the economic restructuring in Mexico and the free trade policies put into place by NAFTA.

Many of the ranchers who belong to the Malpai Borderlands Group trace their roots in the region back five generations. They live in the San Bernadino valley between the Peloncillo and Chiricahua Mountain Ranges. Once these ranges sheltered the Chiricahua Apache, and it was here that in 1886 Geronimo surrendered formally to the U.S. military after being hounded to the point that his people were starving. Since that time, the San Bernadino Valley has remained relatively unpopulated, forming an extensive, natural corridor that favors the north-south movement of wildlife from the largest of predators to the smallest of birds. The few people who live here, for the most part, are ranchers. Their mostly Texan and Kansan forebears, dislocated by the Civil War and the effects of drought in the southern plains, migrated to the stirrup-high grasses surrounding the low, rolling Peloncillo Mountains and the castellated Chiricahua Mountains. They were lured to the desert grasslands by an 1881 offer of free land from the Southern Pacific Railroad and by a burgeoning market for livestock, which was created by the growing number of Indian reservations, military bases, and mining camps of the West. But by the 1920s, it was becoming clear that ranchers could no longer follow in the footsteps of their fathers and grandfathers. The finite boundaries of nature's wealth had begun to appear, as dense grasses gave way to the mesquite trees, cholla cactus, and tall, spiny ocotillo plants that land managers often refer to as invading "aliens."

The ecological deterioration of Southwestern American grasslands led to a vociferous debate over causes and remedies that has raged for over 70 years. Ranchers, environmentalists claim, have caused this deterioration by overgrazing both public and private lands, which has led to erosion, disruption of delicate ecosystems, the decline or extinction of unique species, and the introduction of "exotic" or "alien species." Environmentalists, ranchers counter, rest their case on scientific theories that cannot be proven, and have not invested their own time, sweat, and personal resources in a beloved piece of family-owned ground, nor spent the better part of their lives becoming intimately familiar with every rock, tree, covey of quail, and creek on that land as they move livestock from pasture to pasture. Both ranchers and environmentalists, in turn, place part of the blame for the deteriorating grasslands on

U.S. federal resource managers for not doing enough to rid the world of cows, for doing too much to rid the world of cows, or, even more damning, for suppressing fire in Western American lands for the last 80 years.

Federal resource managers take their own positions in this debate. For much of this century, note Sally Fairfax and Lynn Huntsinger, the U.S. Forest Service and the Bureau of Land Management transformed the basic contours of early ecologist Frederick Clement's theories of succession and progression toward a state of equilibrium or—in the management lexicon, "climax"—into a simple story in which the heroes were the climax plants and the bad guys were the weedy upstarts at the beginning of the path of succession and disturbance. Fire, one of the most common disturbances in the West, was cast as a scary villain that interrupts orderly progression, destroying valuable stands of trees—which the Forest Service considered "climax" species and was charged to protect (Fairfax and Huntsinger 202–203). They argued that educated government and scientific "experts" were needed to disabuse indigenous or rural ranching and farming folk of superstitious and unscientific practices, such as the once-common practice among American Indians and nineteenth-century American ranchers and farmers of burning pastures to encourage the growth of new grass.

This century-long battle between ranchers, government agencies, and environmentalists has fueled a fierce debate over who gets to speak for "the wilderness" and determine what constitutes "proper" management of the still vast tracks of open land in the West. Many ranchers and farmers have taken the position that the decisions are theirs to make, since much of the land at the center of the debate is privately owned. Some have organized a small but growing movement of ranchers and farmers who claim that property rights are absolute and inviolable, and that federal agencies and international environmental organizations like the Nature Conservancy and the Sierra Club do not have the right to make decisions about what is best for nature, life, and the economy of the West. Environmentalists, for their part, often cite these conservative groups as yet further evidence that rural folk are uneducated, unscientific, reactionary, and insensitive to the delicate balance of nature.

In 1993, in response to this ongoing debate, Warner Glenn, owner of the Malpai Ranch, and several of his ranching neighbors formed a non-profit organization called the Malpai Borderlands Group. What surprised many in both the environmentalist and ranching communities is that this new organization openly consulted with and/or sought alliances with some of the very organizations with whom ranchers have historically been in conflict: the Nature Conservancy, the U.S. Forest Service, the Bureau of Land Management, and a team of university conservation biologists and scientists. Even more surprising, the Malpai Borderlands Group takes a position that contrasts with groups claiming that individual property rights are absolute.

In a joint legal agreement, Glenn and his neighbors created something of a commons area or communal lands area by placing conservation easements on their lands—held by the Malpai Group—which prevent them from ever selling to developers intent on breaking their ranches into smaller "ranchettes" or building subdivisions. Thus, they have pledged to keep the over one million acres of public and private land under their joint care, open, unfragmented, and undeveloped.[10]

To grasp how this arrangement protects the members of the Malpai Group economically and, as a result, shields the wild lands and species for which they take responsibility, we must understand something of the economic and environmental challenges that face small ranchers in the borderlands region. The group was formed in the early 1990s, at a time when both drought and passage of NAFTA combined to put a number of ranchers out of business. Conditions had been generally dry since the beginning of the decade but a strong La Niña intensified drought conditions and the economic ripple effect was felt all across the border region. The livestock industry had already been under strain for years as beef prices declined with the shift in people's meat preferences. With the passage of NAFTA—which reduced tariffs on trade in live cattle and beef products—and the intensification of the drought, Mexican livestock flooded U.S. border state cattle auctions. According to Eakin and Liverman, Mexican cattle exports to the United States rose by 50 percent in 1995 and cattle prices plummeted by up to $40 per head as large beef companies bought up the cheaper beef. These combined forces drove a number of ranchers out of business, especially small family-owned enterprises with less than 50 head of cattle (Eakin and Liverman 3).

The ranchers I have spoken to complain about large beef companies, especially the meat packers, which buy beef for the lowest possible prices.[11] These days, much of the cheap beef is imported through Mexico so the producers can take advantage of reduced tariffs under NAFTA. The agreement thus helps to keep the profit of U.S. ranchers low, despite the rising costs of vehicles, equipment, and feed, which ranchers need to keep their ranches operational. This is an especially severe problem for small ranchers, who have little financial margin in the face of the vagaries of the environment, global economic policies, or corporate bottom lines.

Like the small farmers in Chiapas, small ranchers in the Southwest are finding it harder to survive in a world driven by corporate objectives. For them, the creation of an alliance of ranchers makes good economic sense. In return for the conservation easements, they cede to the Malpai Borderlands Group, members of the group share resources with each other in ways that can help them survive the ups and downs of the weather and the market. For example, if one rancher receives too little rain and is facing an economic crisis because of lack of feed, while another rancher has been the recipient of the Southwest's fickle, uneven rains, and the resulting abundant growth of grasses, then the two ranches might agree to share the grasses. This kind of resource sharing helps protect group members from losing their land during bad years.

Malpai ranchers are well aware that they live in a world where the measure of land is its productivity and where the only viable definition of productivity is, usually, a short-term economic one. They argue that unless there is some financial return from the land, they will lose it. The sale and division of their land would not only force ranchers to move to the cities and find other means of supporting themselves, but it would also lead to a loss of biodiversity. Development would prevent wildlife from roaming free through the open spaces between Mexico's Sierra Madre Range and the Peloncillo Mountains. While ecologically degraded grasslands can be restored, the land—and the non-human species that depend on this land for survival—can never recover when it is subdivided and paved over.

Like the Zapatistas, Malpai ranchers believe in their right to protect and control their common lands. This group is entering the struggle over who gets to define "nature" and to determine the future of the region in which their families have lived and worked for many generations. The Malpai refer to the arid grasslands under their joint care as a "working wilderness," a term coined by their neighbor, photographer and author Jay Dusard. In the afterword to Dan Dagget's *Beyond the Rangeland Conflict,* Dusard observes that in arguing for a "working wilderness," ranchers are not rearticulating the dominant Judeo-Christian view that man hold dominion over all other creatures on the earth. Rather, a working wilderness "sustain[s] wildlife and ecosystems" and "livestock and other productive uses" (Dagget 103). The notion of working wilderness directly confronts mainstream U.S. environmentalist discourses that banish humans from nature. Whereas many environmentalist groups argue that world ecological health depends on preserving "global commons areas" in a "pristine condition," these ranchers argue that humans have been part of the ecosystems of the western hemisphere for at least 11,000 years and perhaps more than 30,000. Like the Zapatistas, members of the Malpai Group are arguing that communal lands should be set aside, not simply to preserve the biodiversity of nonhuman species, but also so that humans have a place to work for their own livelihood.

The notion that humans have a right to derive their living from the natural world and to protect their livelihoods in ways that are responsible and sustainable is exemplified in a story that Warner Glenn tells about his encounter with a rare Mexican jaguar that, like its ancestors, traveled up through the Sierra Madre Range to the Peloncillo Mountains. In 1996, Glenn, who is famed in his community as a mountain lion hunter, was tracking what he thought was a large mountain lion on the New Mexican side of the Peloncillos when his dogs bayed the jaguar instead of the cat he thought he was hunting. Glenn knew there had been no confirmed sightings of a Mexican jaguar in the United States since the turn of the last century. He published his photos and a narrative of the event in a small book called *Eyes of Fire: Encounter with a Borderlands Jaguar.* By letting the jaguar go free and donating a portion of the proceeds of the book to a fund to protect the jaguar and its habitat, Glenn confronts long-standing stereotypes about demonic, predator-killing ranchers.

Many ranchers now recognize the role of predators as crucial to the maintenance of healthy ecosystems. Unlike their fathers and grandfathers, they no longer hunt and kill every wild animal they encounter. But far from being an endangered species in the southern Arizona/New Mexico region, the mountain lion is, in fact, steadily increasing in number. Since its growth is accompanied by an increase in lion predation on livestock, expert lion hunters like Glenn are perceived to be performing a valuable service to ranchers who want to protect their livestock and to keep their land economically viable. Glenn's hunts, which he conducts only if it can be confirmed that a particular lion is feeding on livestock, are seen as both a traditional activity and a necessary service to the community. By engaging in responsible hunting, Glenn proclaims that his community has a right to derive a living from the "working wilderness," but that they must do so in ways that conserve and protect the renewable but limited natural resources on which the ranching way of life depends.

Hunting and grazing on public lands has often been demonized in conventional environmentalist rhetoric, largely because of the ranching industry's ruthless and nearly successful efforts to wipe out predators in the western United States. My collection of the oral histories and folklore of the community, however, complicates this view. Community members report that Glenn knows the location of every covey of quail on his land, every Swainson's hawk nest, every pack of coyotes, and that he carefully manages his land so that livestock and wild animals coexist in a mutually beneficial balance. Glenn is a rancher, but he knows the ecology of his region more intimately than most scientists and ecologists, and he practices conservation of its natural resources so that he and his community might continue to enjoy the beauty and bounty of the land. In his narrative, he repeatedly refers to the jaguar as "beautiful," mentions the "surprise and beauty of the scene," and exults in the great luck of seeing a rare jaguar in the vast, open spaces that were once part of its territory (8–9). The jaguar's return to the Peloncillos, argues Glenn, is an indication that the Malpai Group's efforts to restore the ranges surrounding the Peloncillos and to manage for both economic productivity and biodiversity are having a positive effect.

Members of the Malpai Group insist that the knowledge they have gained over five generations in the arid grasslands must be taken into account in any environmental or economic decision making for their region. However, they recognize, just as the Mayan *eijido* farmers who joined the Zapatista National Liberation Army, that the knowledges of peasants or small farmers and ranchers are rarely afforded a place at the table in negotiations over the future of a region. So, like the Chiapas farmers, the Malpai Group has formed alliances with people and groups from outside their immediate community who can help support their interests.

Within the ranching community itself, the Malpai Group's alliance with the Nature Conservancy has been very controversial. Over the last 20 years, the Nature Conservancy has been viewed with great suspicion by many ranchers because of its support of the Endangered Species Act, which has been used in the courts repeatedly by environmentalists to tie up public and private grazing lands in endless court battles. However, the Nature Conservancy has taken notice of several scientific studies, showing that taking cattle off the land does not necessarily improve the range and that excluding all types of disturbance (grazing, fire, trampling) can have a detrimental effect on both the indigenous flora and fauna. They have also begun to realize, as Conservancy Vice President John Cook puts it, "that park-preserve-style efforts aren't going to be enough to close the gap on the care of areas worthy of protection over the next 100 years" (qtd. in Daggett 17). Indeed, Conservancy President John Sawhill asserts that saving uniquely diverse lands "must involve local citizens to be successful" (qtd. in Daggett 17). As a result, the Nature Conservancy is increasingly becoming involved with ranchers and other worker-in-nature groups who are seeking to derive their livelihoods in healthy, sustainable ways that protect biodiversity.

The alliance between the Malpai Group and the Nature Conservancy complicates oversimplified characterizations of "bad" ranchers and "good" environmentalists that have fueled the battle over management of open lands in the West. So do the Malpai Group's efforts to restore ecologically degraded grasslands. While some environmentalists put the blame for the invasion of the grasslands by "alien" or "woody" species—such as mesquite and ocotillo—solely on ranchers and cows, the Malpai Group allows that grazing has played a role in environmental degradation but argues that both drought and suppression of fire have also led to the spread of "alien" species throughout the grasslands. They are supported in this claim by grassland and fire ecologists and environmental historians. For example, Stephen J. Pyne confirms that much of the devastating increase in desert trees and cacti is not just the result of mismanaged cattle grazing, but also of a long ignorance of indigenous fire practices and, in the twentieth century, both persistent drought and failed remedies put into practice by U.S. government "experts" and scientists.

In *World Fire,* Pyne recounts the details of a rancorous, turn-of-the-twentieth-century debate that shaped U.S. rhetoric on fire suppression for the next 70 years. Mexican and American ranchers, following a centuries old traditional fire practice they called the "The Indian Way" or "Paiute Burning," regularly practiced light burning in the front country and, when a fire broke out in the back country, let it burn. This practice discourages the growth of "alien" mesquite and encourages the growth of "native" grasses. But the newly born Forest Service argued that light burning would leave the torch in the hands of folk practitioners instead of professionally trained foresters. Their argument won the day, and their dominance in ecological debates led to the suppression of fire for the next 80 years.

Today, with mesquite and ocotillo growing everywhere in the Sonoran and Chihuahuan Deserts, and with the increasing disappearance of indigenous grasses, the pendulum of expert opinion has swung the other way. Many believe that controlled fire is key to rangeland health, a solution to the uncontrolled spread of woody, alien invaders. Malpai ranchers have successfully reintroduced fire (with both prescribed burns and natural fires they have let burn) and are seeing dramatic improvements in the health of their rangelands.[12] However, many of the ranchers with whom I have spoken caution that while fire must be reintroduced into the ecosystem of the grasslands, it is not a panacea. Not all plant species respond positively to fire. A complex array of elements in the natural world, such as wind, humidity, type of grasses being burned, and the time that has elapsed since the last fire, must all be taken into account before a decision to "let burn" is made. This is one reason that Malpai ranchers have come together with the Nature Conservancy, university scientists, grassland ecologists, and Forest Service and Bureau of Land Management professionals with whom they consult on all prescribed burns and decisions to "let burn." But Malpai ranchers insist that in all debates about proper management of the land, "expert" opinion must be carefully weighed together with five generations of folk knowledge about the animals and plant species of their region. This means that in all discussions of the future of their region, outside knowledges must be weighed carefully together with local knowledges and both must play a role in decisions about how best to manage the land for sustainability and biodiversity.

Working Wilderness or Sacrifice Zone?

Just at the moment when the Malpai Group is drawing lines of protection around a large tract of working wilderness, increasing numbers of South and Central American immigrants are crossing into the United States and impacting the environment as they follow riverbeds and arroyos and leave tons of garbage along their routes. The infamous Barnett brothers would have us believe that this is a simple story of white, property owning, environmentally concerned "native" ranchers versus landless, brown "aliens" who cross the international border, damage property, and wreak havoc on the environment. When asked what the solution is, Roger Barnett does not hesitate to call for area ranchers to mount vigilante responses. He adds that if the Border Patrol cannot stem the tide of immigrants, "[W]e need to get the military on the border" (qtd. in Skinner 2). This kind of muddled rhetoric taps into popular concern for the environment as a justification for vigilante actions, while it fails to account for the specific ways in which the Border Patrol supports and maintains the political and corporate institutions that write off the border region as sacrificeable to global economic objectives and allow the people who live there—on both sides of the border—to be discounted.

Throughout the 1980s and 1990s, as the United States shifted its concerns from alleged threats from foreign terrorists and drugs to the protection of national territory against the influx of "dangerous" people, particularly those of Hispanic origin. In a special report on the militarization of the border, Carol Nagengast writes that "opinion makers, politicians, and Congress have portrayed the border area and the communities within it as places 'infested' with hordes of drug runners, welfare cheats and foreigners looking for a free ride. The Border Patrol, as an arm of the state, has been charged with keeping 'our' country safe from these scourges" (38).

To this end, the Border Patrol has overseen the construction of walls along the border and set up a series of check points that force border-crossing immigrants away from populous cities such as El Paso and San Diego toward the dangerous, desert terrains of Arizona. Douglas, Arizona, a small border town near the Peloncillo Mountains is now the most frequently used starting point for migration into the United States. Every day from early Spring through late Fall, thousands of immigrants pass through Douglas and attempt to cross the desert, often with only a plastic jug of water, a few clothes, and a small amount of food. Many leave the jugs and other refuse along the way but make it successfully to their destination. Some lose their lives after becoming lost or disoriented in the searing Arizona heat. The Border Patrol's strategy, then, risks harm to the environment by channeling migration through a targeted area while, at the same time, it wields the environment itself as a weapon in the battle to stop illegal migration into the United States.

It could be argued that this tactic transforms working wilderness into a sacrifice zone, a place in which U.S. politicians and the Border Patrol agents they direct claim the right to speak for and make decisions about the future of the region and the people who live there. Those left out of the decision-making process are the people who inhabit the region. Certainly, the ranchers living near the Peloncillos were not called into the deliberations over transforming their region into a barrier against immigration or consulted as to the impacts such a strategy would have on indigenous flora and fauna, livestock, the environment, or their economic livelihood.

I have already discussed the adverse economic consequences of passage of NAFTA on U.S. ranchers and the ways in which the economic restructuring in Mexico has contributed to increased migration. But when we factor NAFTA into discussions of the Border Patrol's role in the conflict between ranchers and undocumented immigrants, it also becomes possible to see that the sacrifice zone on the border is transnational and that U.S. ranchers are not the only group of "sacrifice people." As Timothy Dunn writes in *The Militarization of the U.S.-Mexico Border,* economic integration and border militarization are not necessarily mutually exclusive developments. Dunn notes the correlation between an expansion in U.S. investments in Mexico and the militarization of the border. He writes that,

> the buildup of border-region immigration and even overlapping drug enforcement efforts may have served to intimidate and discourage some would-be undocumented immigrants. In effect, this would have constituted a form of "disciplining" workers by encouraging them to remain in their home countries, where wages were low (at least on a global scale) and general labor rights were restricted. (159)

Such a situation is advantageous to both domestic and international investors in Mexico, as it helps keep wages down.

Mexico's politicians and the chief executive officers of multinational corporations that have located their factories in Mexico tout the wages earned by workers employed in the *maquiladoras* scattered all along the border as the highest in the nation. But a drive through the *colonias,* or squatter villages, surrounding every *maquiladora* in Agua Prieta, Sonora, or Nogales, Sonora, quickly reveals that these "highly paid" workers are living in houses built from cardboard and wooden pallets that lack basic sanitation facilities and are located in places without access to clean water. People living in the *colonias* are forced to have their water delivered by tankers, which draw from chemically contaminated wells located underneath industrial sections of town. The delivered water is then stored in chemically contaminated barrels scavenged from the *maquilas.*[13]

As Claudia Sadowski-Smith discusses in more detail in her contribution to this collection, in the sense that it "disciplines" workers and attempts to keep them on the low-paying side of the international border, the Border Patrol supports and helps maintain multinational businesses in Mexico that have benefited from NAFTA. The Border Patrol bases their "disciplinary" tactics on a doctrine of "low intensity conflict" (LIC), which is fueled by the same dualistic, native-versus-alien rhetoric spouted by the Barnett brothers. The essence of LIC doctrine is the establishment and maintenance of social control over targeted civilian populations through a broad range of sophisticated measures by police, paramilitary, and military forces. One of the doctrine's distinguishing characteristics is that military forces take on police functions throughout a specifically targeted region (Dunn 4).[14]

Rather than reify old dualism, the Malpai Borderlands Group's careful use of fire in their management of "indigenous" grasses and its involvement in the protection of the Mexican jaguar raises all sorts of questions about easy distinctions between different species or groups of people. For example, if the jaguar has been missing from an ecosystem for over 80 years and then reappears in the region, is it "native" or "alien"? If drought, which is usually considered to be a natural phenomenon, plays a role in the spread of "alien" species such as the mesquite, then is mesquite "alien" or "native" when it spreads throughout the grasslands? If human-caused fires have played a role in the evolution of the grasslands over millenia, then are human-caused fires "natural" or "unnatural"? If we learn anything from Warner Glenn's encounter with the jaguar, the Malpai Group's about-face on the issue of predators (which they now see as key to the balance of healthy ecosystems) and the Malpai Group's careful study of how complicated a matter it is to decide what is "indigenous" or "exotic," it is that overly simplistic "good/bad," "native/alien" constructions often give us permission to ignore the specificity of particular problems in particular places or ignore elements of a problem that need additional or more complex courses of action. Certainly, to reduce the conflict between ranchers and immigrants on the U.S.-Mexico border to a simple story of "natives" and "aliens" is to ignore the larger economic and environmental forces at work in the region.

The Malpai Group has not weighed in on the issue of immigration with any kind of official statement, but if we learn anything from their careful management of "working wilderness" and their protection of the Mexican jaguar—which many indigenous Mesoamericans still associate with shamanic power and the past glory of Mayan and Aztecan royalty and which, as a consequence, is a fitting symbol of the current conflict between "natives" and "aliens" on the border—it is that we must question easy dualistic constructions that describe the distinction between different species or different groups of people. Moreover, by forming an alliance that crosses the once seemingly unbridgeable chasms between ranchers, federal agents, environmentalists, and scientists, the Malpai provide us with a model for how differing groups of people might come together to address the mounting social, economic, and environmental problems we face in the twenty-first century. This group is just one of the many that are mobilizing all over the world to respond to threats to their cultural and economic survival and to the sacrifice of the environments in which they live. Such grassroots mobilization is as dramatic as the Zapatistas' rebellion or as surprising as a ranching group organizing to protect a rare jaguar. These two groups' actions transcend depoliticizing dualisms to reveal larger global forces at work on the local level. The Malpai experience suggests that while solving complex problems will not be easy, transformative changes can only happen when all interested parties are called into the deliberations over the future of particular places.

Notes

1. See "Arizona Vigilantes Threaten" and "Arizona Vigilantes Shed More Mexican Blood."
2. See the work of Inger Sandal and of Ignacio Ibarra.
3. In interviews with local ranchers, I have been taken to see fences that have been cut, sometimes again and again on well-traveled migration routes, or which have been run through with vehicles. This exposes cattle to injury or death, as it allows them to escape onto roads where they can be hit by vehicles or lost to the environment or predators. Float valves on water tanks have been broken, leaving water to run on the ground, and ranchers are subsequently forced to truck in water to their cattle, something that is both a time and a financial hardship.
4. See the work of Daniel Dombey and of Homero Aridjis.
5. Ibid.
6. During the colonial period, the 90 percent decline in indigenous population led to the diminished protection of common lands, but with Mexican independence and a series of laws aimed at promoting individual property titling and the disentailment of lands held by the Church, Mexico's indigenous people began losing their remaining common lands at a rapid pace. By 1910, Mexico's indigenous population was deprived of 90 percent of its land (Stephen 76).
7. Highly conscious of international opinion and the media attention focused on their impoverished state, the Mexican government was constrained from using the full force of its military to put down this rebellion—as it had others in the history of Chiapas. They tried to discredit the rebels by alleging that the indigenous members of the army were being duped by "radical groups" or "professionals" who were "tricking and even forcing the participation of Indians."
8. Already throughout the 1960s and 1970s, the people of Chiapas had seen that discoveries of petroleum deposits in their state did little to improve their lot. Where oil was found, communal lands were expropriated and profits from the drilling were directed away from Chiapas toward the more prosperous center of Mexico, leaving the people of Chiapas, again, without their lands and without basic infrastructure to make their *eijidos* competitive in an increasing global market (Golden A16).

9. I take the notion of a "sacrifice zone" from the Nixon administration, which, in 1972, proposed designating several American Indian reservations and Western American desert landscapes (which had already been toxically contaminated by nuclear weapons testing and uranium and coal mining operations) as "National Sacrifice Areas" where the government and multinational corporations would be free to dispose of toxic wastes. The people who inhabited these areas, thus would become "national sacrifice peoples," and their voices, knowledges, and material needs would be discounted. I discuss the politics of creating "national sacrifice zones," at more length in *American Indian Literatures, Environmental Justice, and Ecocriticism: The Middle Place.* See especially chapters Two and Three of that work, which focus on the Four Corners region of the American Southwest.
10. For more on the early organization and goals of the Malpai Borderlands Group, see Verlyn Klinkenborg and Jake Page.
11. For more on the adversarial relationship between ranchers and the meat packing industry, see Larry Aylward.
12. See "The Gray Ranch," in Dan Dagget, 13–23. The Gray Ranch is one of the largest ranches in the Southwest—500 square miles—and is at the heart of the Malpai Group's reintroduction of fire into the grasslands. Dagget notes that "lightning-started fires left to burn naturally [have] revitalized nearly 123,000 acres of the Gray and other ranches of the Borderlands area in 1993 and 1994" (22).
13. On July 17, 2000, I accompanied Teresa Leal, a resident of Nogales, Arizona, and co-chair of the Southwest Network for Environmental and Economic Justice on a tour of the squatter villages of Nogales, Sonora.
14. Anyone driving from the Douglas, Arizona area, where I live and teach, to Tucson, Arizona, or Nogales, Sonora, knows that at various points along the route you will be stopped at a Border Patrol checkpoint, and asked if you are a U.S. citizen. If you have Anglo features and a nice car, you will be waved through by the friendly, smiling agents. However, if you happen to be of Hispano or Mexicano descent, you will be questioned more rigorously, and perhaps asked to step out of the car, even if you are a U.S. citizen. This has the effect of coopting U.S. "natives" (or those *thought* to be "natives") who see the friendly, helpful agents everyday and become used to the idea of them policing the region, into the Border Patrol's mission to stop the flow of illegal "aliens." "Natives" help Border Patrol agents by patiently putting up with the minor inconvenience of a traffic stop. While a large majority of people who cross the border are neither drug runners nor terrorists (but often people, such as farmers from Chiapas who can no longer support themselves by farming), the Border Patrol often makes a show at its check points by keeping large buses filled with captured "aliens" waiting, while the "natives" are waived through. I have also seen the Border Patrol post signs at these check points indicating the number of "aliens" caught that day or month. These tactics have the effect of policing the distinction between "natives" and "aliens" and criminalizing people of His-

panic and Mexican descent on both sides of the border by reducing a host of complicated issues to questions of race.

Works Cited

Adamson, Joni. *American Indian Literature, Environmental Justice, and Ecocriticism: The Middle Place.* Tucson, AZ: University of Arizona Press, 2001.

Aylward, Larry. "Unrest on the Range," *Meat Market and Technology* (October 1998): 36–46.

Aridjis, Homero. "Chiapas Revolt Rooted in a Repressive History," *Arizona Daily Star* (January 10, 1994): A13.

"Arizona Vigilantes Threaten Peace at the Border: Dangerous Flash Point Developing" and "Arizona Vigilantes Shed More Mexican Blood at the Border," *La Voz de Aztlán* 1.10 (May 8, 2000) Available at www.aztlan.net. Accessed May 21, 2000.

Dagget, Dan. *Beyond the Rangeland Conflict: Towards a West that Works.* With portraits by Jay Dusard. Layton, UT: Gibbs Smith, The Grand Canyon Trust, copublisher, 1995.

Dombey, Daniel. "Chiapas Uprising Surprises a Mexico in Denial: History of Unrest Is Overlooked as Part of Government Snubbing," *Arizona Daily Star* (January 1, 1994): A1 and A16.

Dunn, Timothy. *The Militarization of the U.S.-Mexico Border, 1978–1992.* Austin, TX: CMAS Books, 1996.

Eakins, Hallie and Diana Liverman. "Drought and Ranching in Arizona: A Case of Vulnerability," *Impact of Climate Change on Society.* USGS Webpage (July 10, 1997): 1–9. Available at http://geochange.er.usgs.gov/sw/impacts/society/ranching. Accessed June 7, 2000.

Fairfax, Sally and Lynn Huntsinger. "An Essay from the Woods (and Rangelands)," *Arizona Quarterly* 53.2 (Summer 1997): 191–210.

Glenn, Warner. *Eyes of Fire: Encounter with a Borderland Jaguar.* El Paso, TX: Printing Corner Press, 1996.

Golden, Tim. "Peasant Groups, Church Members Accused of Rebel Aid," *Arizona Daily Star* (January 9, 1994): A16.

Goldman, Michael." Introduction: The Political Resurgence of the Commons," *Privatizing Nature: Political Struggles for the Global Commons.* Ed. Michael Goldman. London: Pluto Press, 1998: 1–19.

Ibarra, Ignacio. "Barnetts Had An Earlier Problem, Records Show," *Arizona Daily Star* (July 15, 2000): B4.

Klinkenborg, Verlyn. "Crossing Borders," *Audubon* (September-October 1995): 35–47.

Nagengast, Carol. "Militarizing the Border Patrol," *NACLA: Report on the Americas.* Special Report on Militarization. 32.3 (November/December 1998): 37–41.

Page, Jake. "Finding Common Ground in the Range War," *Smithsonian* 28.3 (July 1997): 50–61.

Pyne, Stephan J. *Fire in America: A Cultural History of Wildland and Rural Fire.* (1982) Seattle, WA: University of Washington Press, 1997.

———. *World Fire: The Culture of Fire on Earth.* New York: Henry Holt and Company, 1995.

Sandal, Inger. "Jury Orders Towing Firm to Pay $1.3M," *Arizona Daily Star* (July 15, 2000): B1.

Simon, Joel. "Zapatistas Caution Against Election Fraud," *Arizona Daily Star* (August 8, 1994): A11.

Skinner, Ed. "A Human Tragedy Unfolds: Illegal Immigrants Leaving Debris, Damage, Disease," *Sierra Vista News* (August 3, 2000): 1–2 and 16.

Stephen, Lynn. "Between NAFTA and Zapata: Responses to Restructuring the Commons in Chiapas and Oaxaca, Mexico," *Privatizing Nature: Political Struggles for the Global Commons.* Ed. Michael Goldman. London: Pluto Press. 1998: 76–101.

Notes on Contributors

JONI ADAMSON is Associate Professor and Program Head of English and Folklore at the University of Arizona, Sierra Vista, where she teaches American Literature, Environmental Literature, and Folklore. She is the author of *The Middle Place: Native American Literature, Environmental Justice, and Ecocriticism* (University of Arizona, 2001).

URSULA BIEMANN studied art and critical theory in Mexico and at the School of Visual Arts and the Whitney Independent Study Program in New York. Her art and curatorial work on gender relations in economy, media, and urbanity rearticulates notions of postcoloniality and visual representation. Some of her recent activities include collaborative exhibition projects, publications with migrant women in Switzerland, and *Kültür,* a two-year project on gender and urban politics in Istanbul. Involved since 1988 with the U.S.-Mexico border, in 1999 Biemann produced a video essay, entitled *Performing the Border;* followed in 2000 by *Writing Desire,* a video on female sexuality and the bride market in cyberspace; and in 2001 by *Remote Sensing,* a video on the geography of the global sex trade. Her most recent art publication is *been there and back to nowhere: gender in transnational spaces* (Berlin: b_books, 2000).

ARLENE DÁVILA teaches Anthropology and American studies at New York University. She is the author of *Latinos Inc: The Marketing and Making of a People* (University of California Press, 2001) and *Sponsored Identities: Cultural Politics in Puerto Rico* (Temple University Press, 1997).

CLAIRE F. FOX is associate professor of English at The University of Iowa. She is the author of *The Fence and the River: Culture and Politics at the U.S.-Mexico Border* (Minnesota University Press, 1999), and her articles have appeared in *Discourse, Social Text, Iris,* and *Studies in Twentieth Century Literature.*

DONALD A. GRINDE, JR. (YAMASEE) is professor of History and director of ALANA/Ethnic Studies at the University of Vermont. He is the co-author of several books, including *Debating Democracy: Native American Legacy of*

Freedom (Clear Light, 1998), the *Encyclopedia of Native American Biography* (Da Capo Press, 1998), and *The Ecocide of Native America* (Clear Light, 1995). Professor Grinde is currently completing a monograph entitled *Human Rights and Home Rule: The Mission Indian Federation in Southern California,* and was recently commissioned by the U.S. Congress and Congressional Quarterly Press to write *A Political History of Native Americans.*

MANUEL RAFAEL MANCILLAS is a community organizer, teacher, and performing artist from the Tijuana/San Diego border region. He is a member of the Border Arts Workshop/Taller de Arte Fronterizo (BAW/TAF) and is currently performing with Puerto Amerexico, a transcultural music and poetry ensemble. Mancillas has written several articles and online pieces, and has been invited to lecture on U.S.-Mexico border issues at the University of Illinois, the California Arts Council in San Francisco, and Illinois State University. Mancillas is currently working on a community organizing project for educational justice with Mexican immigrants in San Diego County.

MANUEL LUIS MARTINEZ is assistant professor of English at Indiana University, Bloomington, where he also works in Chicano and American Studies. He is the author of a novel, *Crossing,* and his study *Countering the Counterculture: Rereading Dissent in Postwar America* is forthcoming from Verso Press. He is currently beginning a project on the social criticism of Ernesto Galarza and has completed a second novel, *Drift.*

SHARON NAVARRO is the 2001–02 Academic Minority Resident Scholar at Carleton College in Northfield, Minnesota. She received a Ford Dissertation Fellow for the year 2000–2001 and recently completed a dissertation on Latinas and cross-border mobilization at the U.S.-Mexico border in the Political Science department at the University of Wisconsin-Madison. Dr. Navarro is the co-author, with C. Richard Bath, of *Elections and Party Politics in the State of Chihuahua: 1982–1994* (Center for Inter-American and Border Studies, 1996).

BRYCE TRAISTER is associate professor of English at the University of Western Ontario, London, Canada. His research interests include eighteenth- and nineteenth-century authorship and masculinity, colonial religious history and gender, and contemporary theories of American literature and culture. He has published articles in these areas in the *Canadian Review of American Studies, American Literature, American Quarterly,* and *Studies in American Fiction.* His first book-length project will be a study of female religious expression and the invention of the American secular, 1630–1800.

CLAUDIA SADOWSKI-SMITH is assistant professor of English at Texas Tech University. She has published articles on border studies, literatures of the U.S.-Mexico border, globalization theory, and post–Wall East-Central Europe in *Diaspora, Arizona Quarterly, Postmodern Culture,* and *The Comparatist.* Currently, she is at work on a book-length project, entitled *Border Fictions: Globalization and Transnationalism in Literature about U.S. Borders.*

Index